DEDICATION

This Whiskey Steak Guns & Freedom Shooting Log is dedicated to all the safety focused and responsible gun owners out there who are looking to improve. We know that getting better at shooting takes work and time on the range and keeping track of your stats and progress plays a big part in getting better!

YOU are my inspiration for producing shooting log books and I'm honored to be a part of keeping all of your essential shooting information, experiences, and notes organized all in one easy to find spot.

Be safe and enjoy!

HOW TO USE

This Whiskey Steak Guns & Freedom shooting Log is the perfect book to improve your shooting! Here are a few pointers on how to use the log:

1. What gets logged/recorded improves!
2. SAFETY First
3. Write or underline any lessons learned. Summarize any lessons learned for easier memory.
4. Trigger safety/discipline is huge.
5. A small amount of practice (dry fire and live fire) every day is better than hitting the range once a year.
6. Share your findings and learn from others of course!

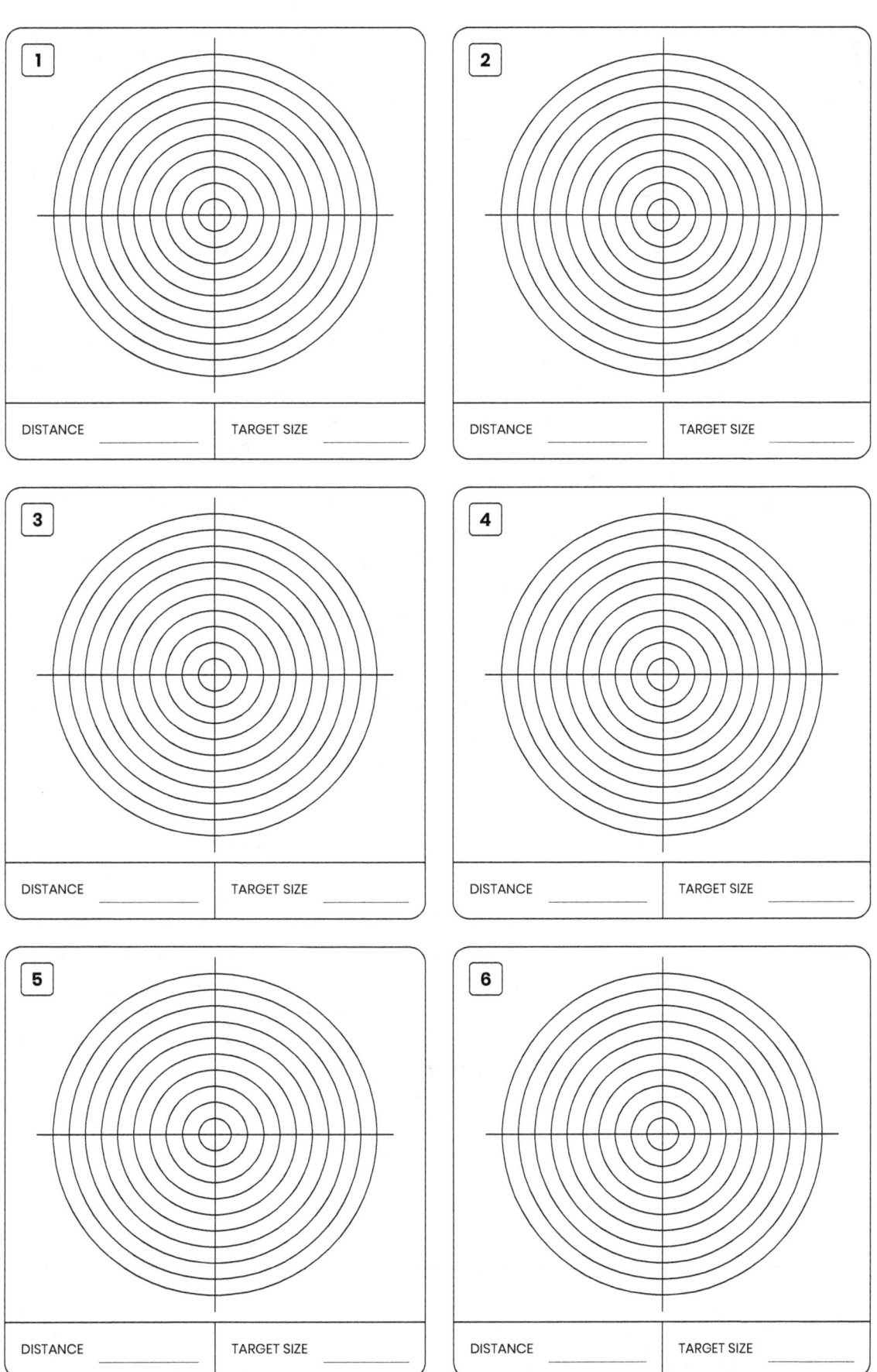

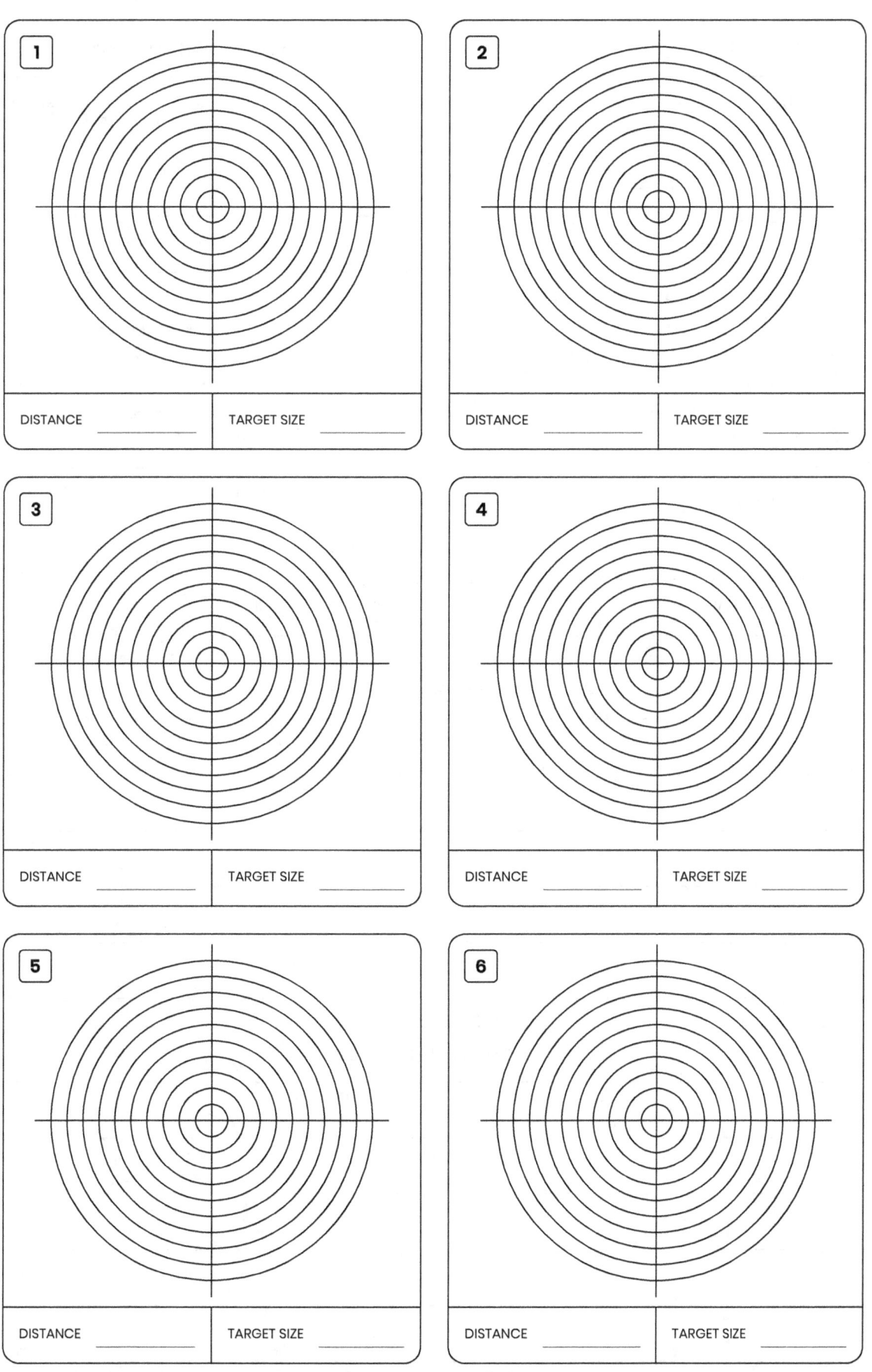

📅 DATE	
🕐 TIME	
📍 LOCATION	
👥 PARTNER	
🔫 FIREARM	
🎯 SCOPE TYPE	
🔸 BULLET TYPE	
POWDER	
PRIMER	
BRASS	
SEATING DEPTH	

WEATHER CONDITIONS

🌡️ _____   ☀️ ⛅ 🌧️ ⛈️ ❄️
🚩 _____   ☐ ☐ ☐ ☐ ☐

LIGHT CONDITIONS

☀️ 1   2   3   4   5   🌙
BRIGHT ○   ○   ○   ○   ○   DARK

RATING

WEAPON HANDLING	☆☆☆☆☆
HIT RATE	☆☆☆☆☆
OVERALL RESULTS	☆☆☆☆☆

SESSION HIGHLIGHTS

_____
_____
_____
_____
_____

ADDITIONAL NOTES

_____
_____
_____
_____
_____

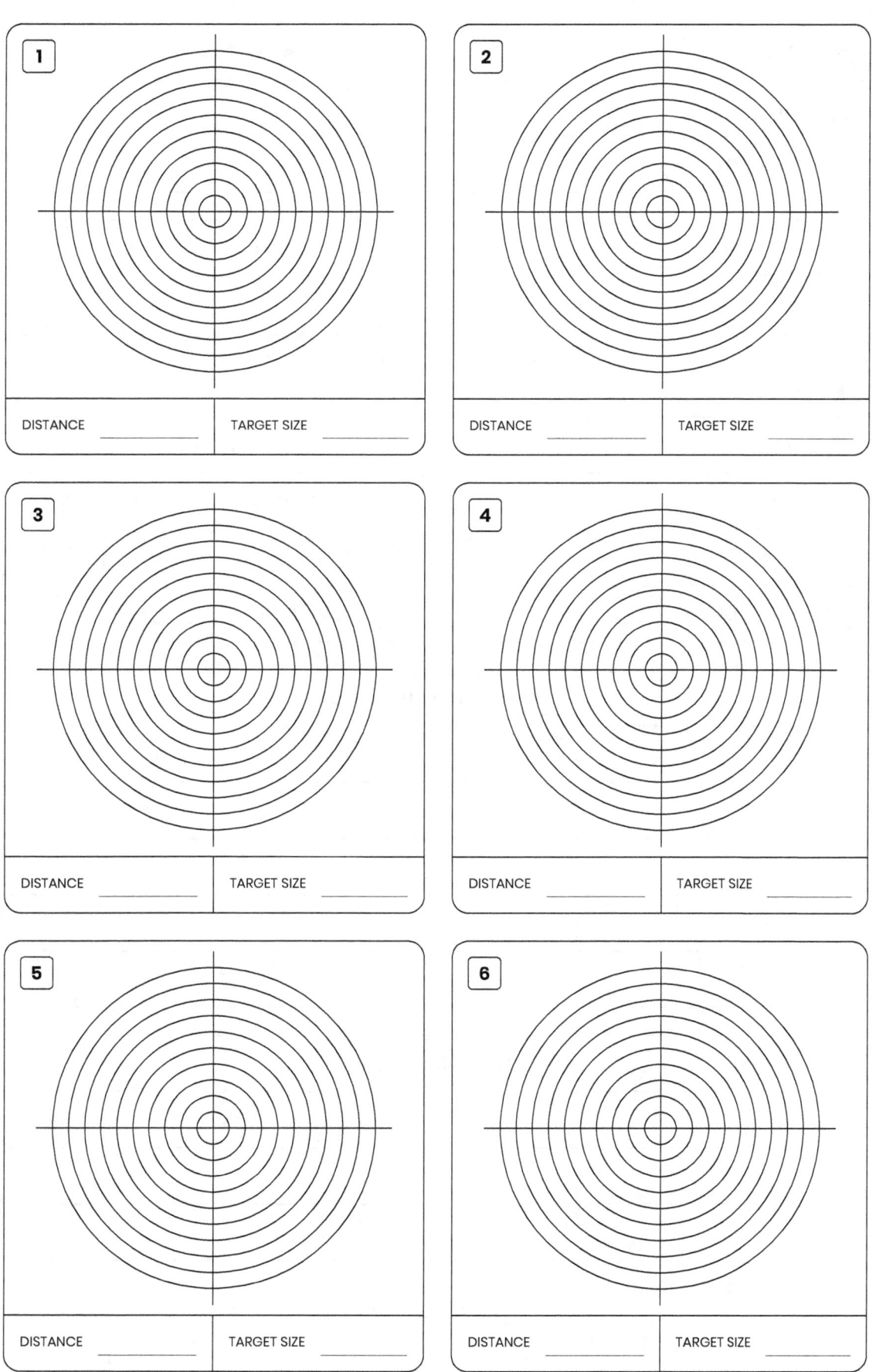

📅 DATE	
🕐 TIME	
📍 LOCATION	
👥 PARTNER	
🔫 FIREARM	
🎯 SCOPE TYPE	
🔹 BULLET TYPE	
POWDER	
PRIMER	
BRASS	
SEATING DEPTH	

WEATHER CONDITIONS

🌡 _____   ☀ ⛅ 🌧 ⛈ ❄
🚩 _____   ☐ ☐ ☐ ☐ ☐

LIGHT CONDITIONS

☀ 1 — 2 — 3 — 4 — 5 🌙
BRIGHT ○ ○ ○ ○ ○ DARK

RATING

🔫	WEAPON HANDLING	☆☆☆☆☆
🎯	HIT RATE	☆☆☆☆☆
📋	OVERALL RESULTS	☆☆☆☆☆

SESSION HIGHLIGHTS

_____
_____
_____
_____
_____

ADDITIONAL NOTES

_____
_____
_____
_____
_____

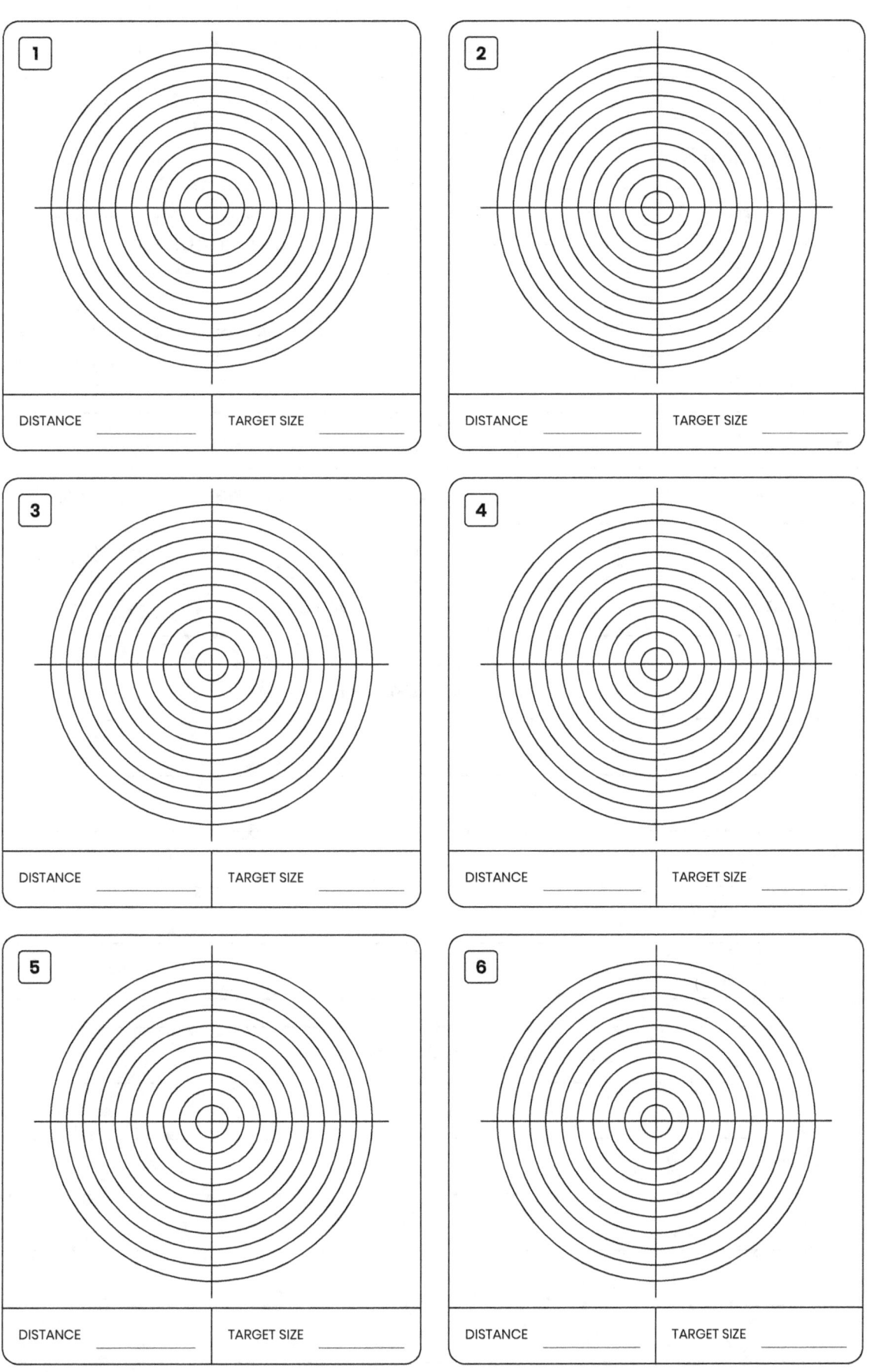

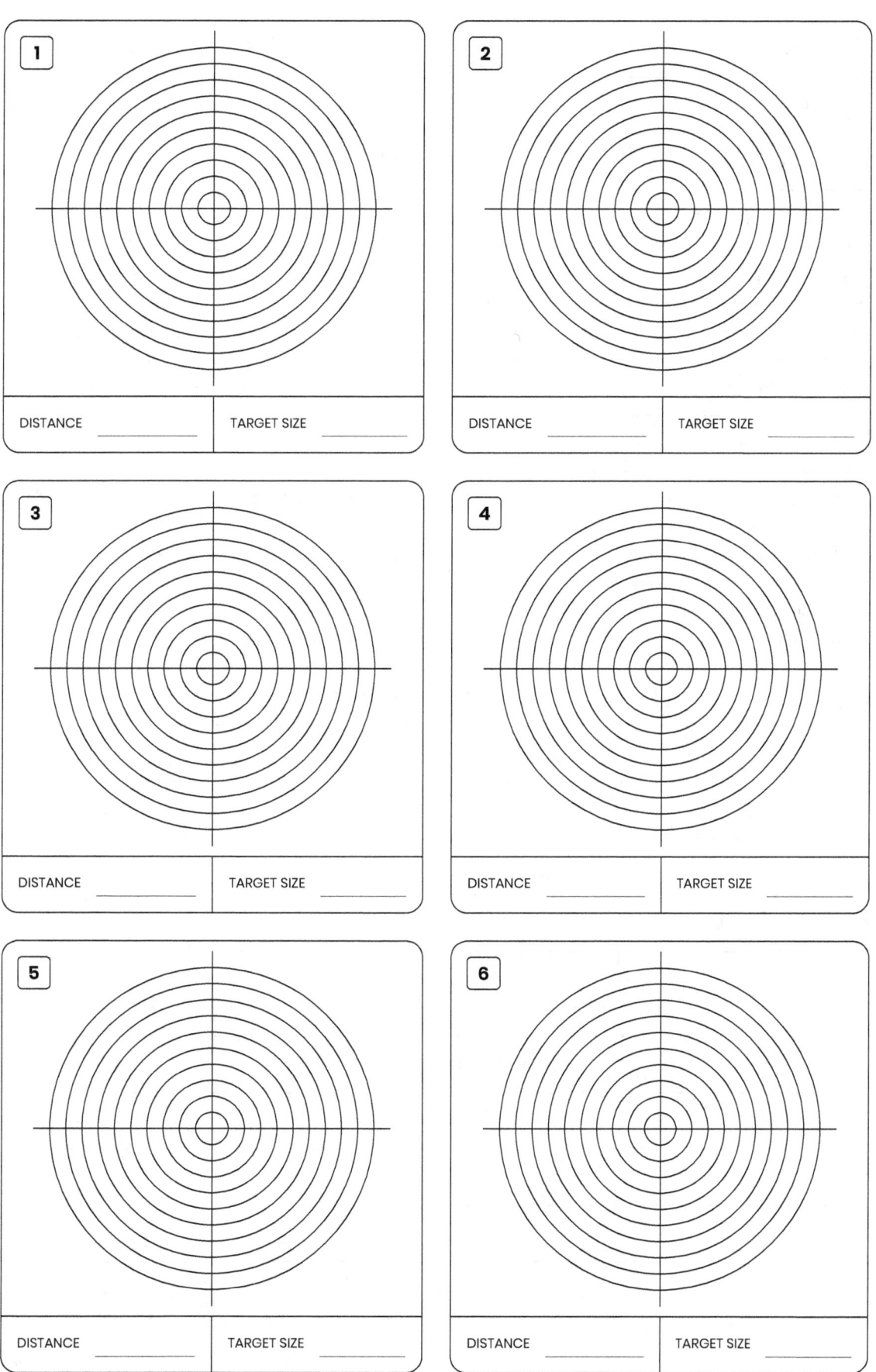

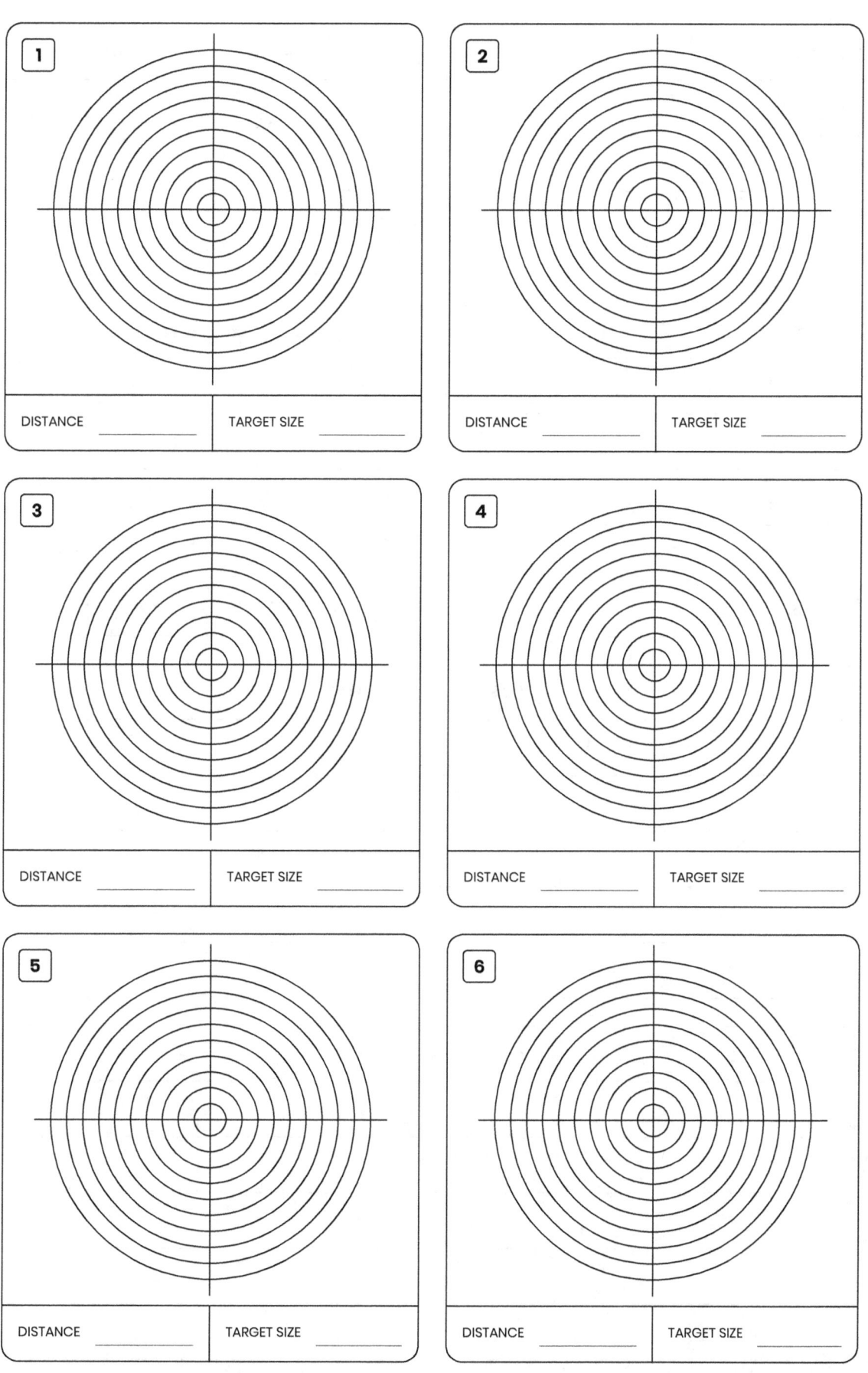

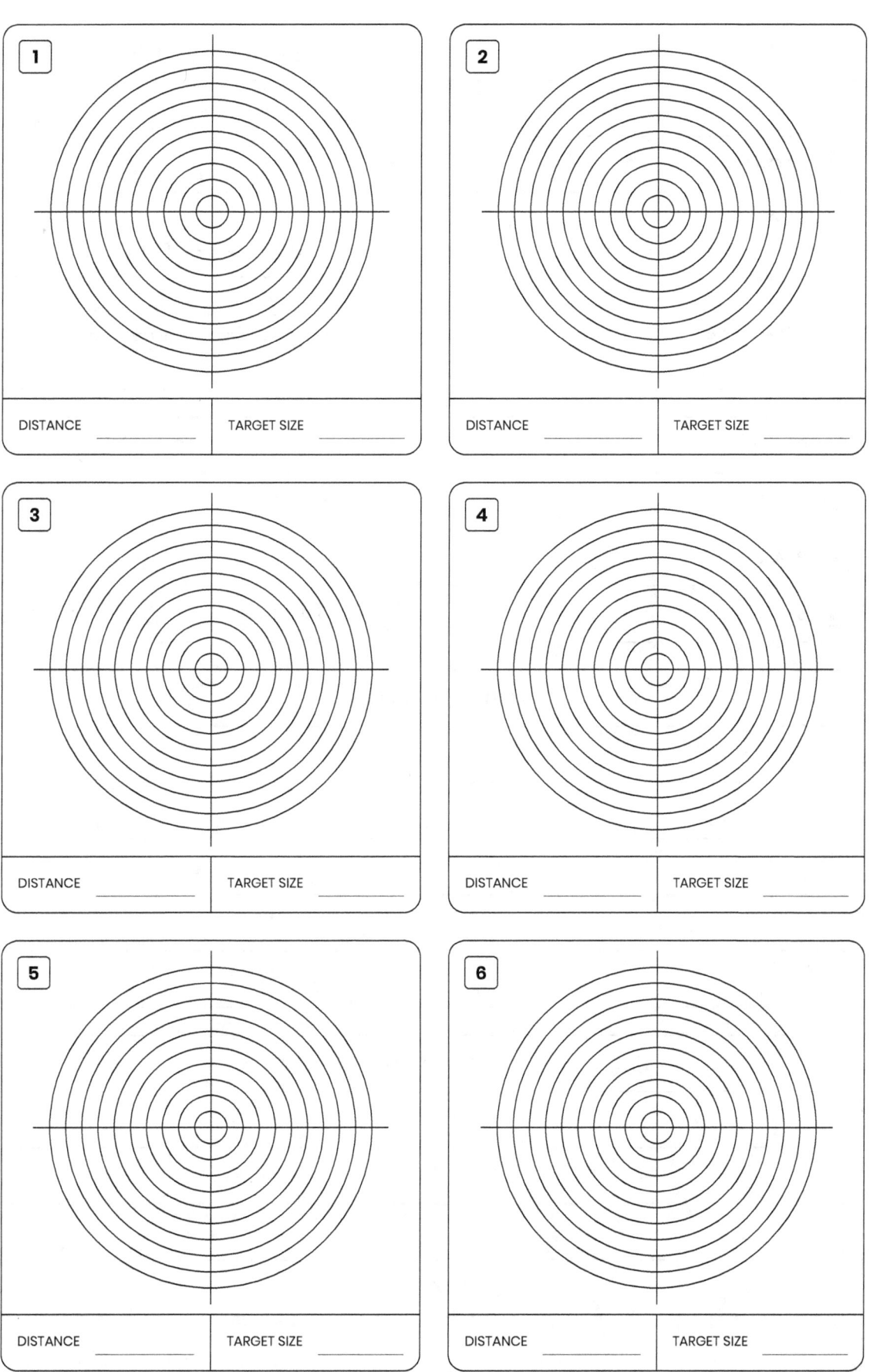

📅	**DATE**
🕐	**TIME**
📍	**LOCATION**
👥	**PARTNER**
🔫	**FIREARM**
⊚	**SCOPE TYPE**
🔹	**BULLET TYPE**
🔺	**POWDER**
💥	**PRIMER**
✏️	**BRASS**
🔫	**SEATING DEPTH**

WEATHER CONDITIONS

🌡 ____   ☀️ ⛅ 🌧 ⛈ ❄️
🚩 ____   ☐ ☐ ☐ ☐ ☐

LIGHT CONDITIONS

☀️ 1 — 2 — 3 — 4 — 5 🌙
BRIGHT ○   ○   ○   ○   ○ DARK

RATING

🔫	WEAPON HANDLING	☆☆☆☆☆
🎯	HIT RATE	☆☆☆☆☆
📋	OVERALL RESULTS	☆☆☆☆☆

SESSION HIGHLIGHTS

___

___

___

___

___

ADDITIONAL NOTES

___

___

___

___

___

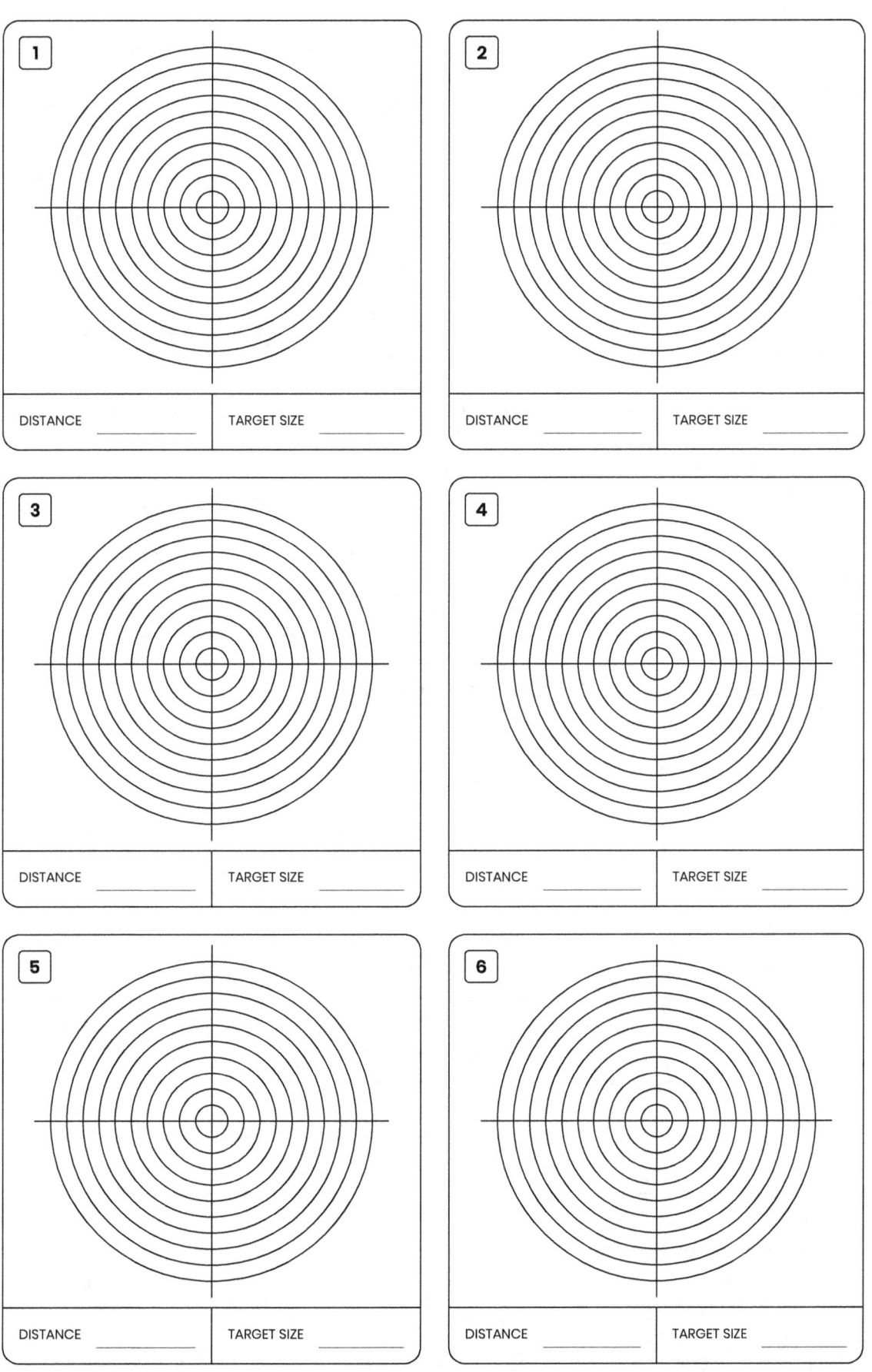

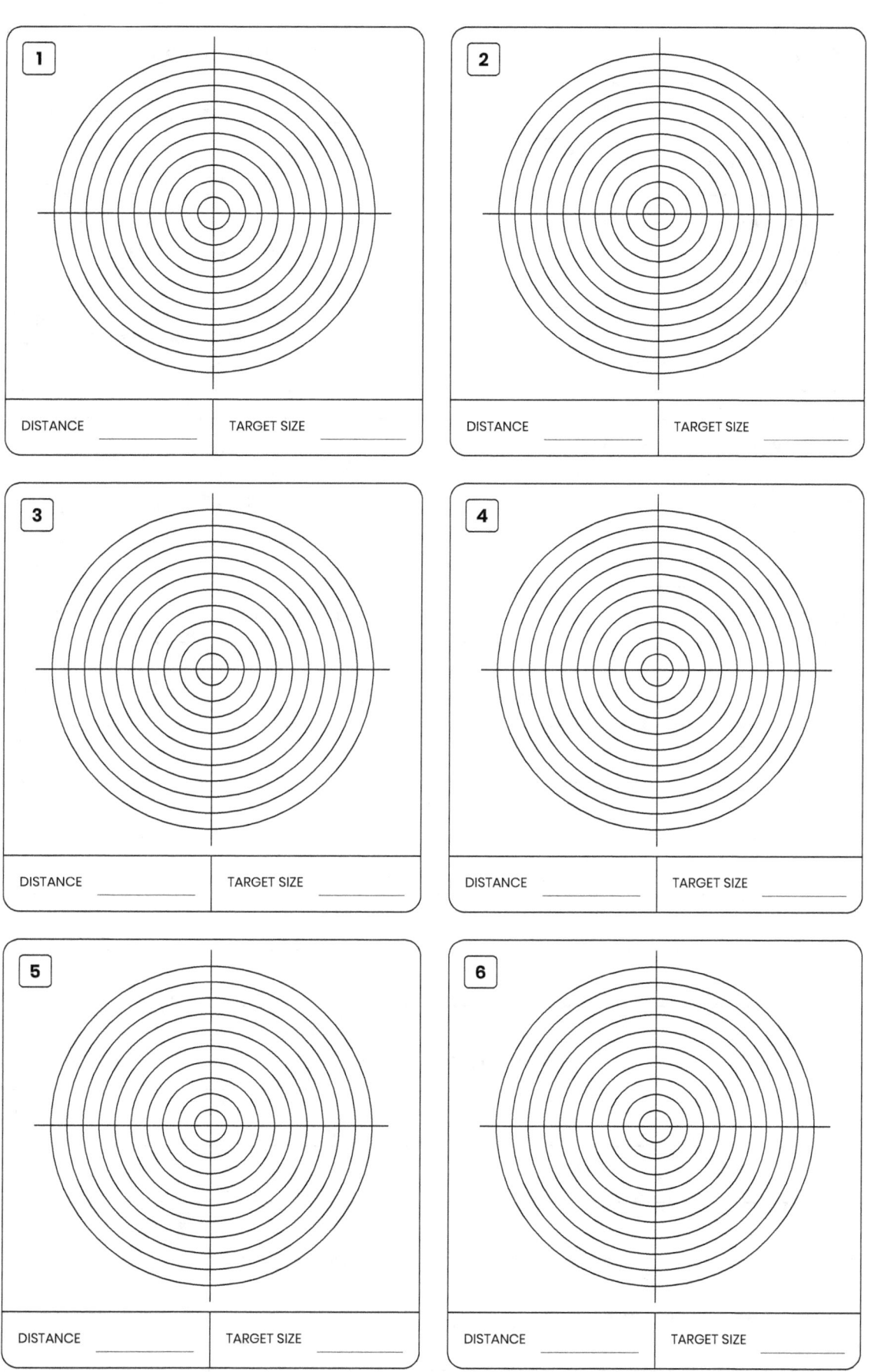

📅	**DATE**
🕐	**TIME**
📍	**LOCATION**
👥	**PARTNER**
🔫	**FIREARM**
🎯	**SCOPE TYPE**
🔸	**BULLET TYPE**
	POWDER
	PRIMER
	BRASS
	SEATING DEPTH

WEATHER CONDITIONS

🌡 _____   ☀️ ⛅ 🌧 ⛈ ❄️
🚩 _____   ☐   ☐   ☐   ☐   ☐

LIGHT CONDITIONS

☀️ BRIGHT — 1 — 2 — 3 — 4 — 5 — 🌙 DARK

RATING

🔫	WEAPON HANDLING	☆☆☆☆☆
🎯	HIT RATE	☆☆☆☆☆
📋	OVERALL RESULTS	☆☆☆☆☆

SESSION HIGHLIGHTS

_____
_____
_____
_____
_____

ADDITIONAL NOTES

_____
_____
_____
_____
_____

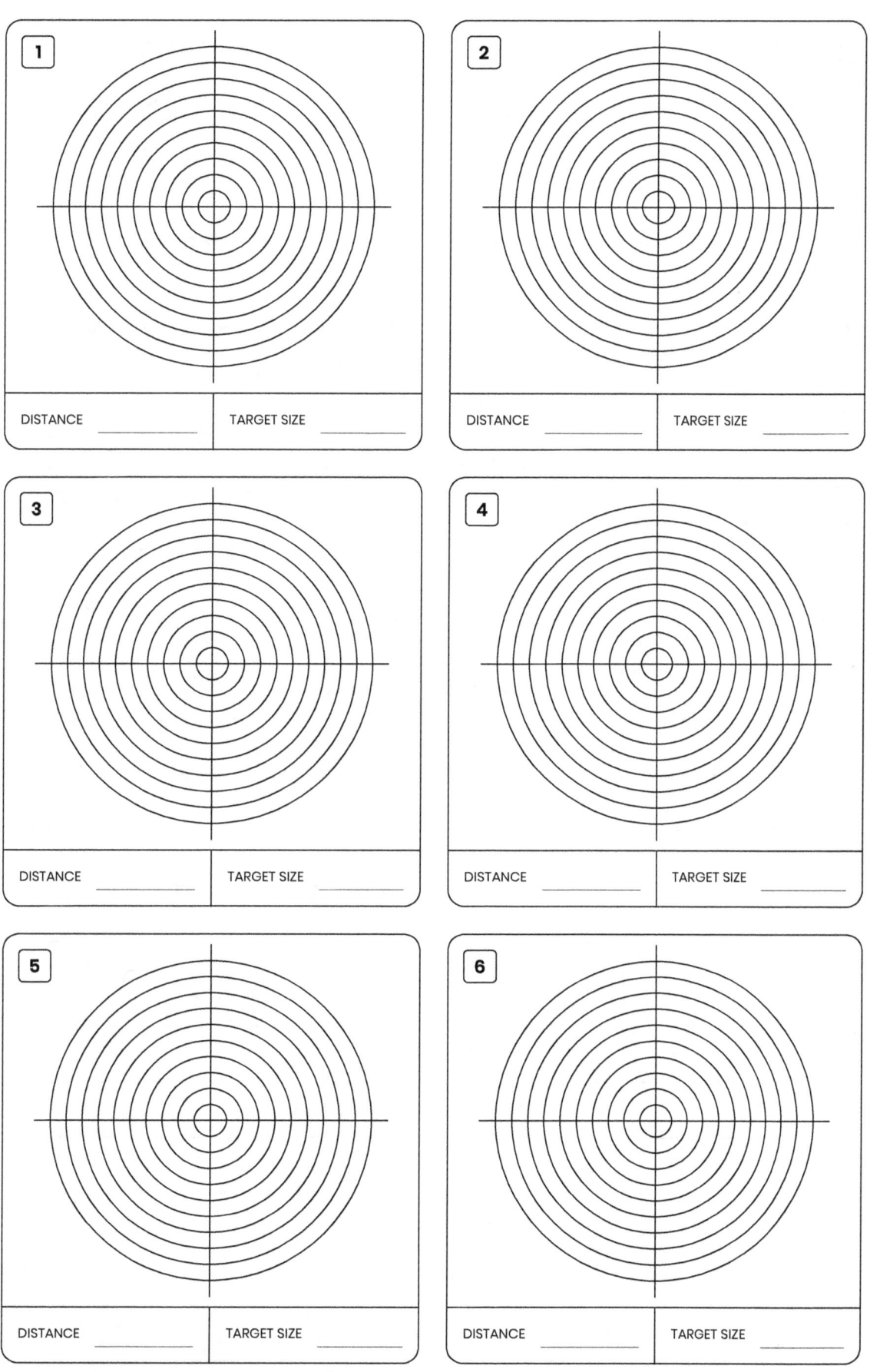

📅 DATE	
🕐 TIME	
📍 LOCATION	
👥 PARTNER	
🔫 FIREARM	
🎯 SCOPE TYPE	
🔸 BULLET TYPE	
POWDER	
PRIMER	
BRASS	
SEATING DEPTH	

WEATHER CONDITIONS

🌡 _____   ☀ ⛅ 🌧 ⛈ ❄
🚩 _____   ☐ ☐ ☐ ☐ ☐

LIGHT CONDITIONS

☀ BRIGHT   1   2   3   4   5   🌙 DARK
             ○   ○   ○   ○   ○

RATING

WEAPON HANDLING		☆☆☆☆☆
HIT RATE		☆☆☆☆☆
OVERALL RESULTS		☆☆☆☆☆

SESSION HIGHLIGHTS

_____
_____
_____
_____
_____

ADDITIONAL NOTES

_____
_____
_____
_____
_____

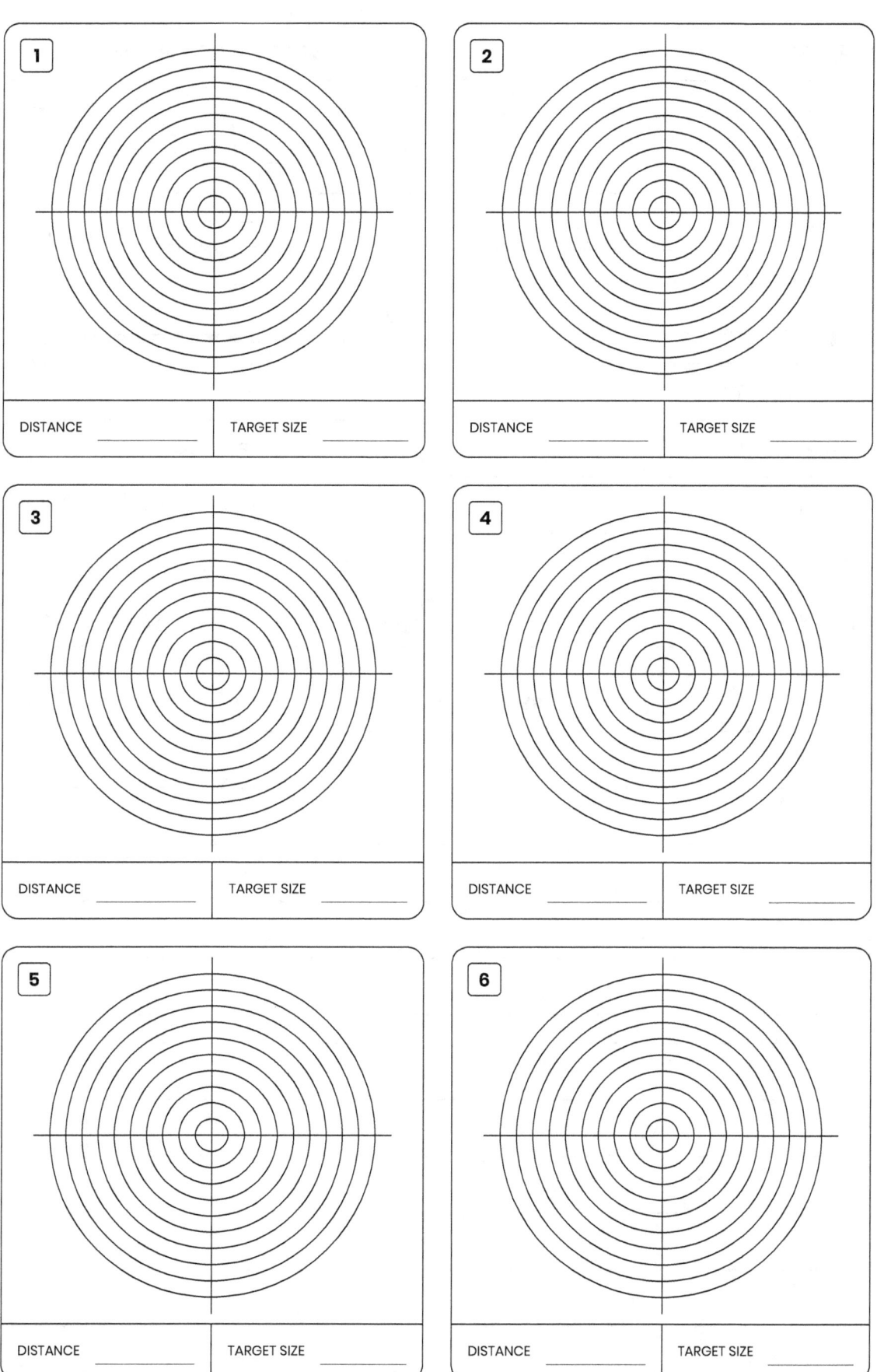

📅	**DATE**
🕐	**TIME**
📍	**LOCATION**
👥	**PARTNER**
🔫	**FIREARM**
◎	**SCOPE TYPE**
🔸	**BULLET TYPE**
	POWDER
	PRIMER
	BRASS
	SEATING DEPTH

WEATHER CONDITIONS

🌡 ____   ☀️ ⛅ 🌧 ⛈ ❄️

🚩 ____   ☐   ☐   ☐   ☐   ☐

LIGHT CONDITIONS

☀️ 1 — 2 — 3 — 4 — 5 🌙
BRIGHT   DARK

RATING

🔫	WEAPON HANDLING	☆☆☆☆☆
🎯	HIT RATE	☆☆☆☆☆
📋	OVERALL RESULTS	☆☆☆☆☆

SESSION HIGHLIGHTS

_____
_____
_____
_____
_____

ADDITIONAL NOTES

_____
_____
_____
_____
_____

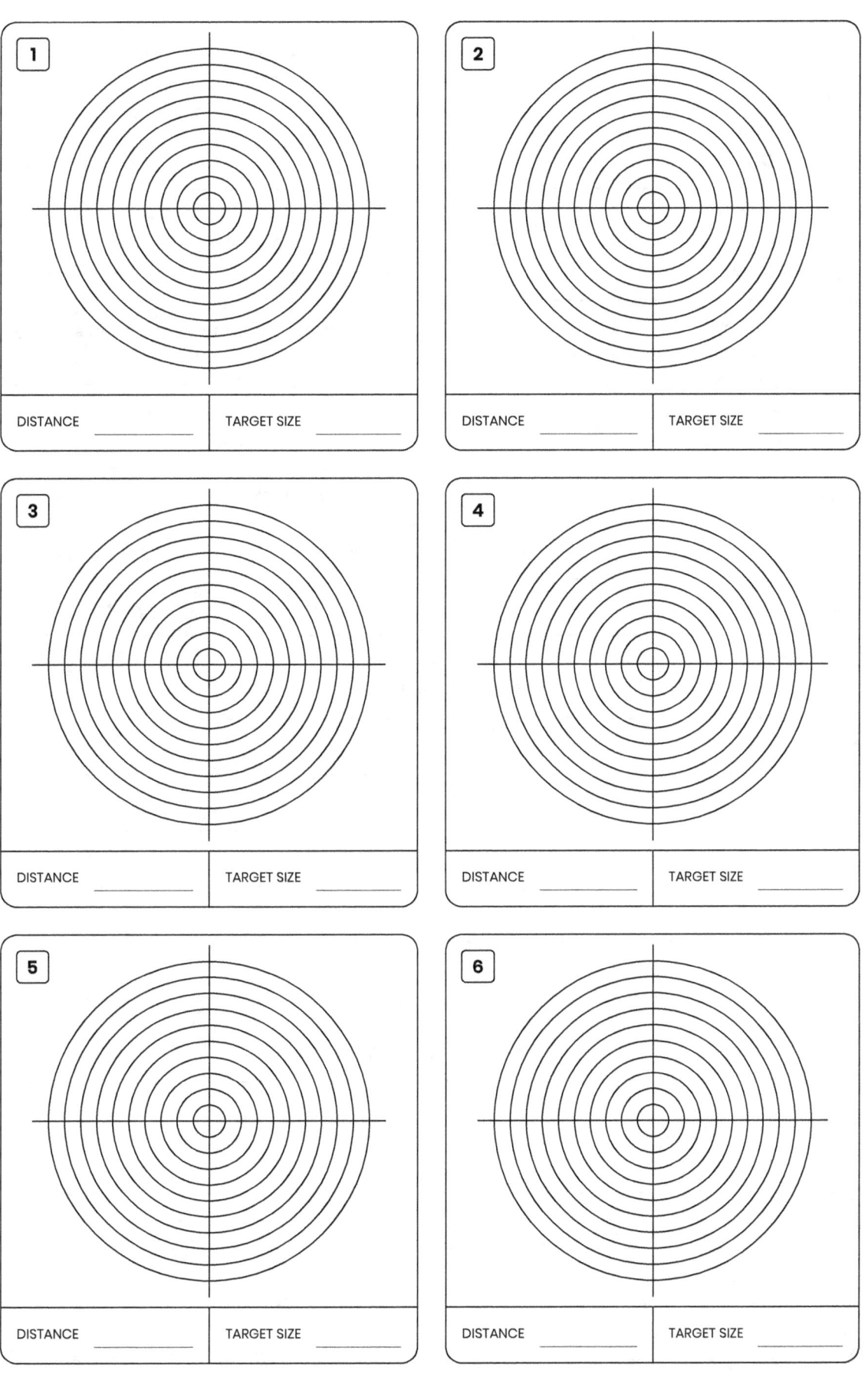

📅	**DATE**
🕐	**TIME**
📍	**LOCATION**
👥	**PARTNER**
🔫	**FIREARM**
🎯	**SCOPE TYPE**
🔹	**BULLET TYPE**
	POWDER
	PRIMER
	BRASS
	SEATING DEPTH

WEATHER CONDITIONS

☀️ ⛅ 🌧️ ⛈️ ❄️

LIGHT CONDITIONS

BRIGHT 1 — 2 — 3 — 4 — 5 DARK

RATING

🔫	WEAPON HANDLING	☆☆☆☆☆
🎯	HIT RATE	☆☆☆☆☆
📋	OVERALL RESULTS	☆☆☆☆☆

SESSION HIGHLIGHTS

ADDITIONAL NOTES

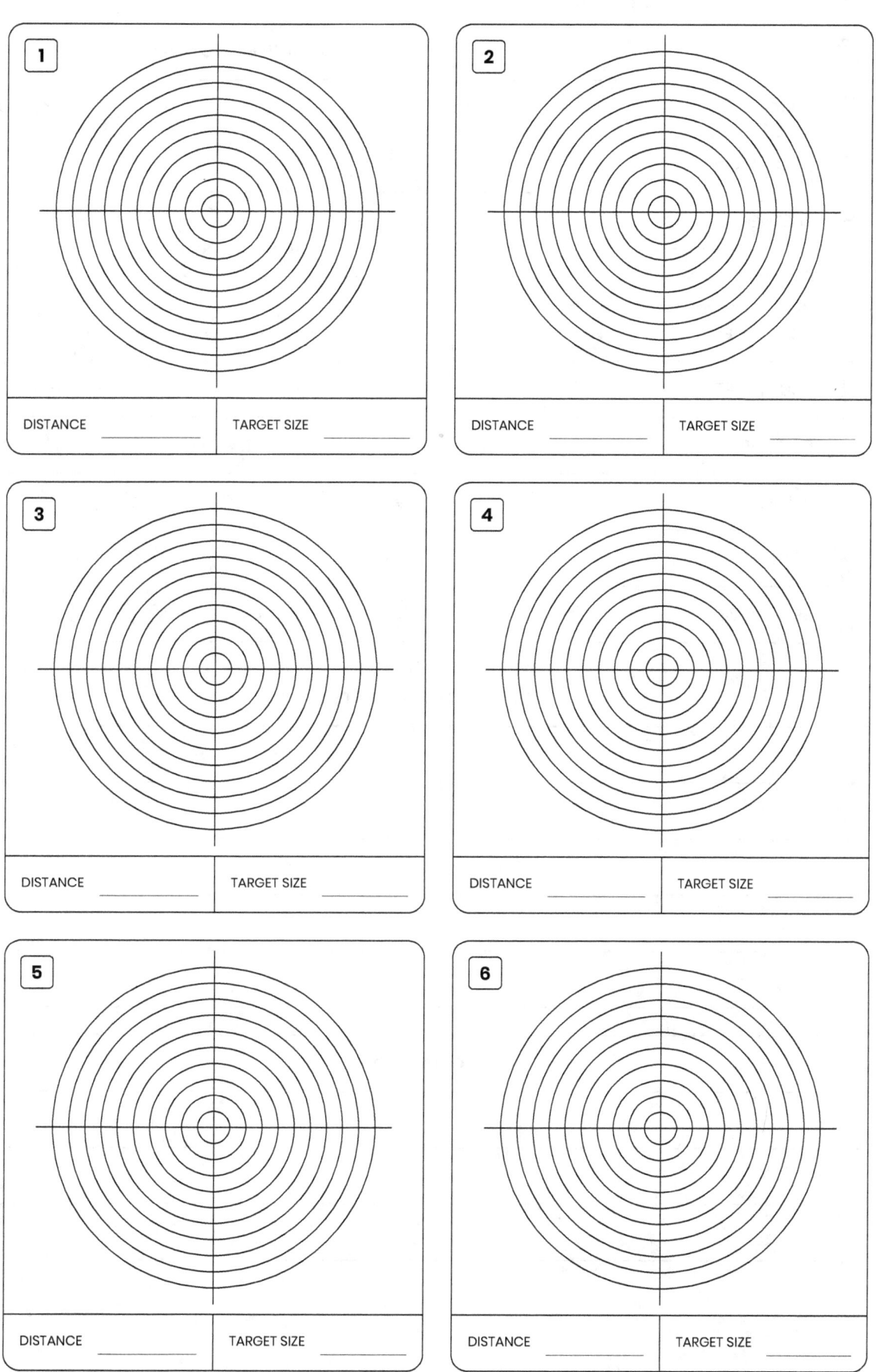

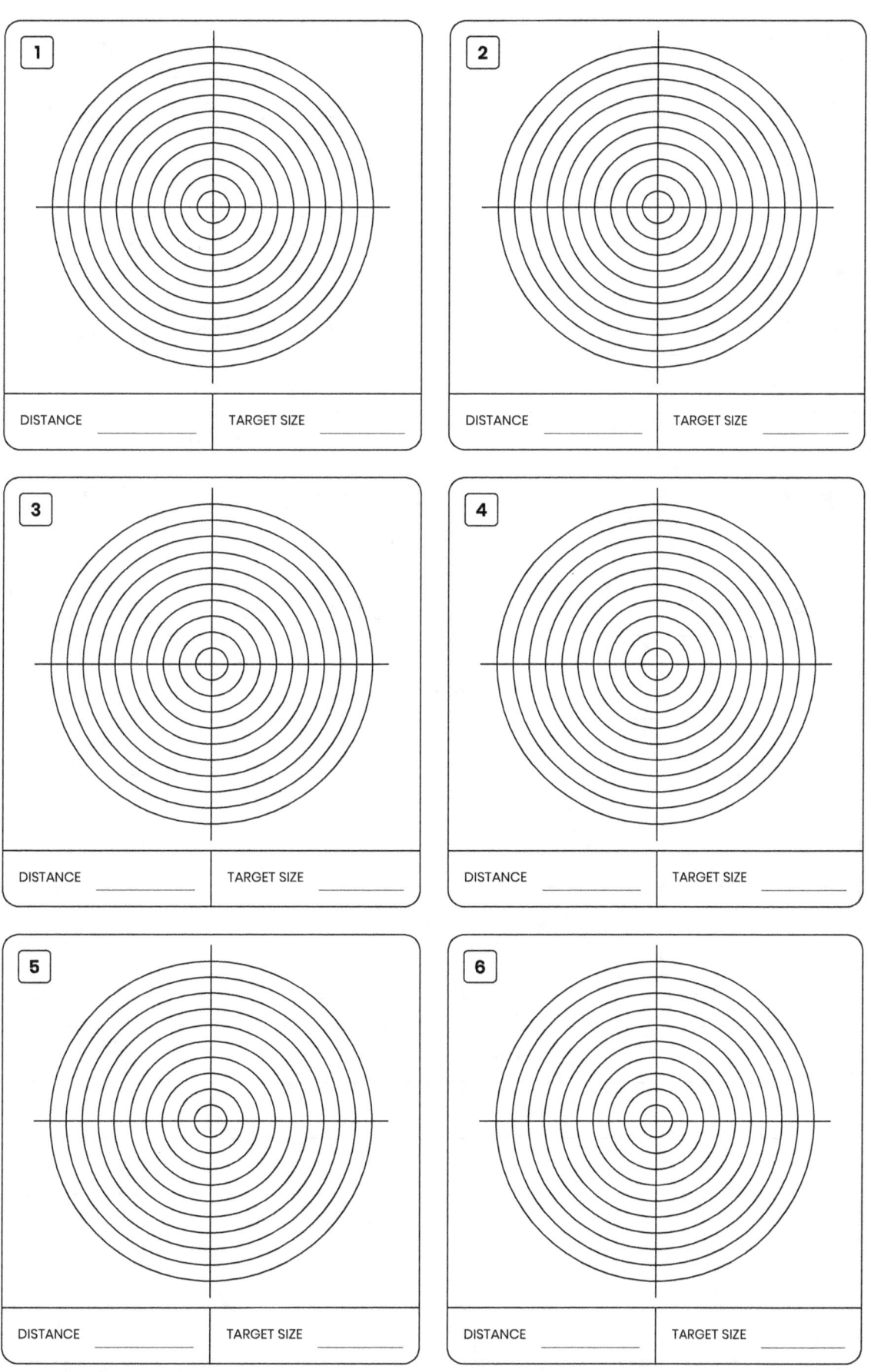

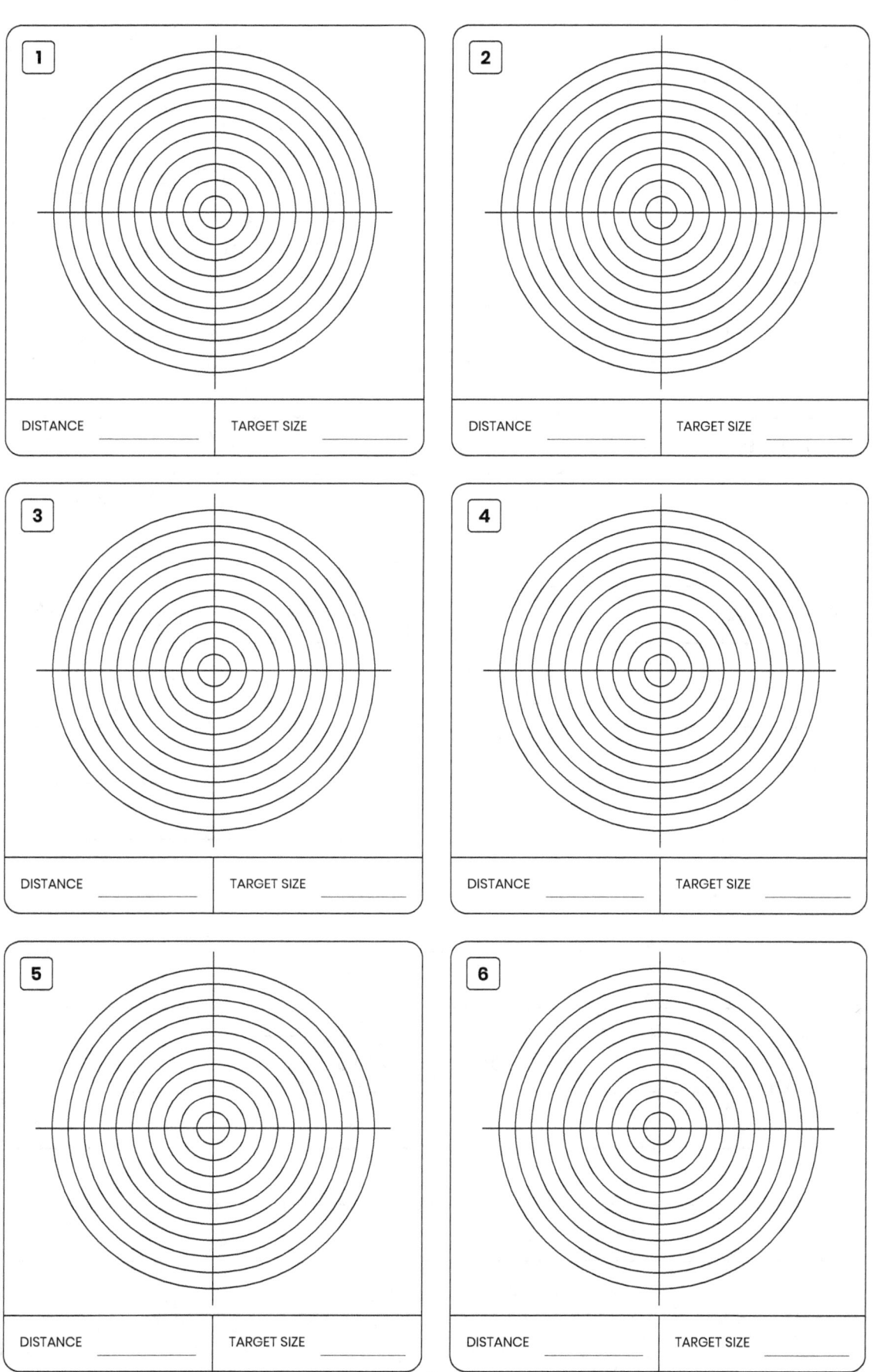

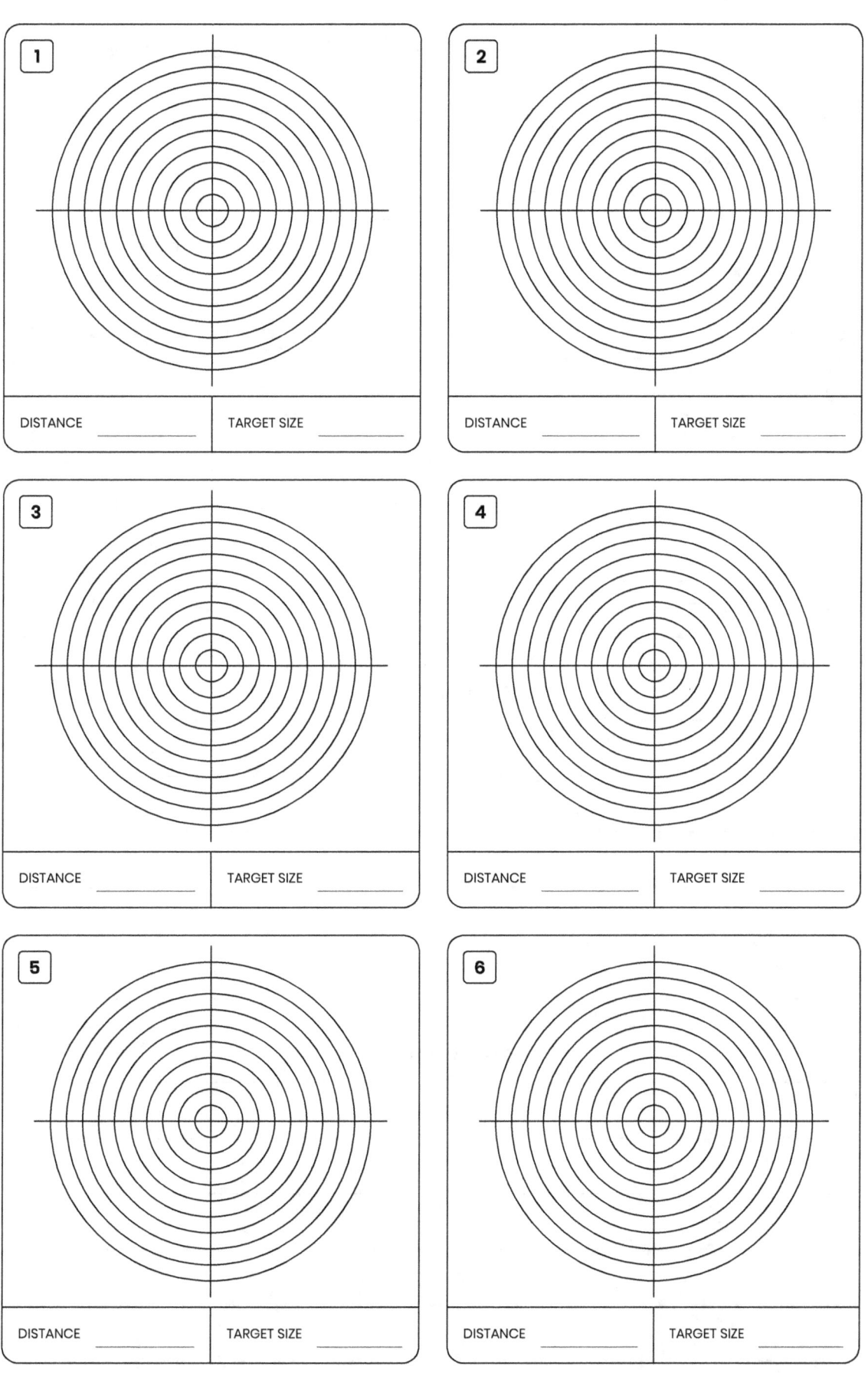

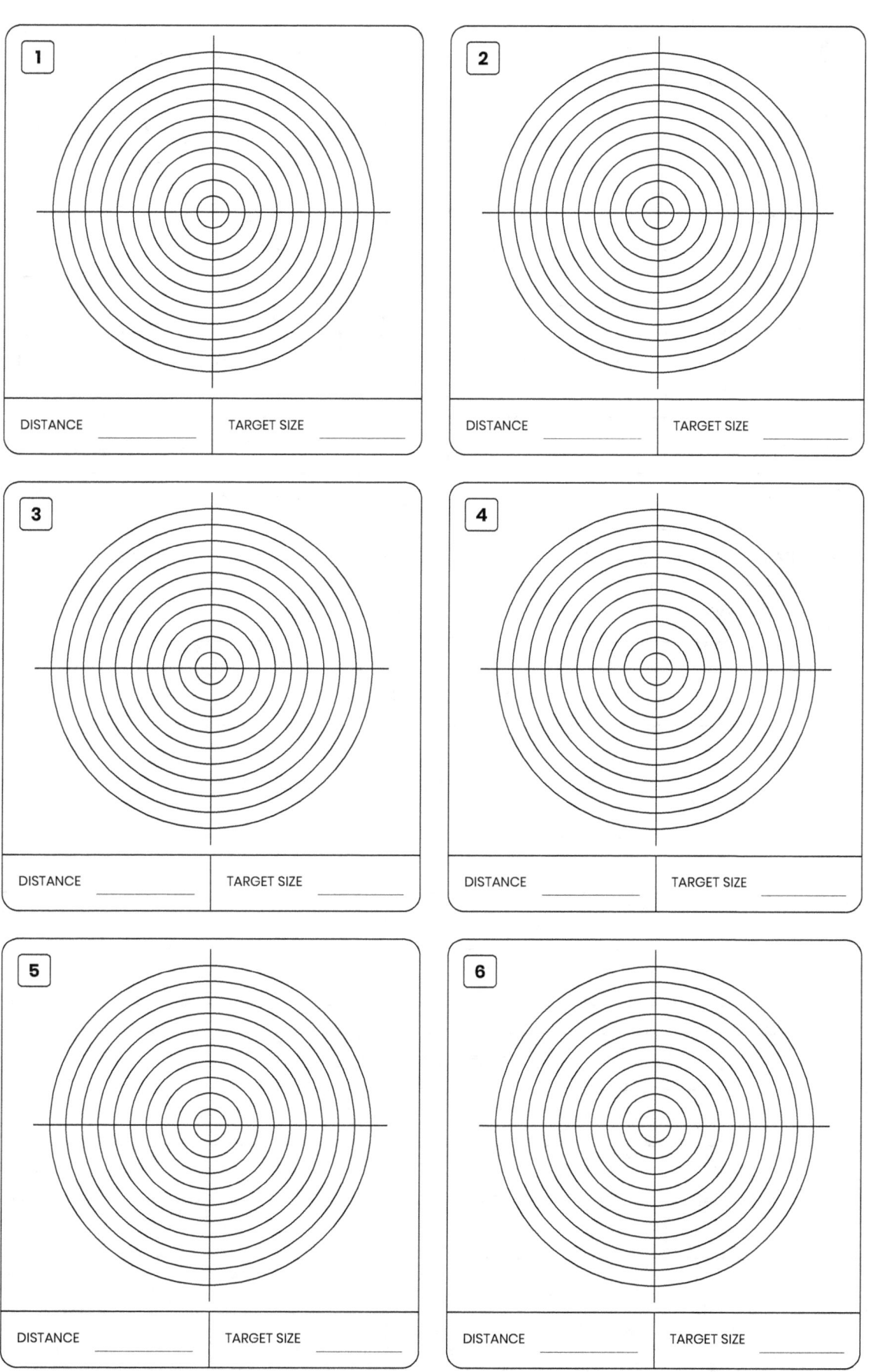

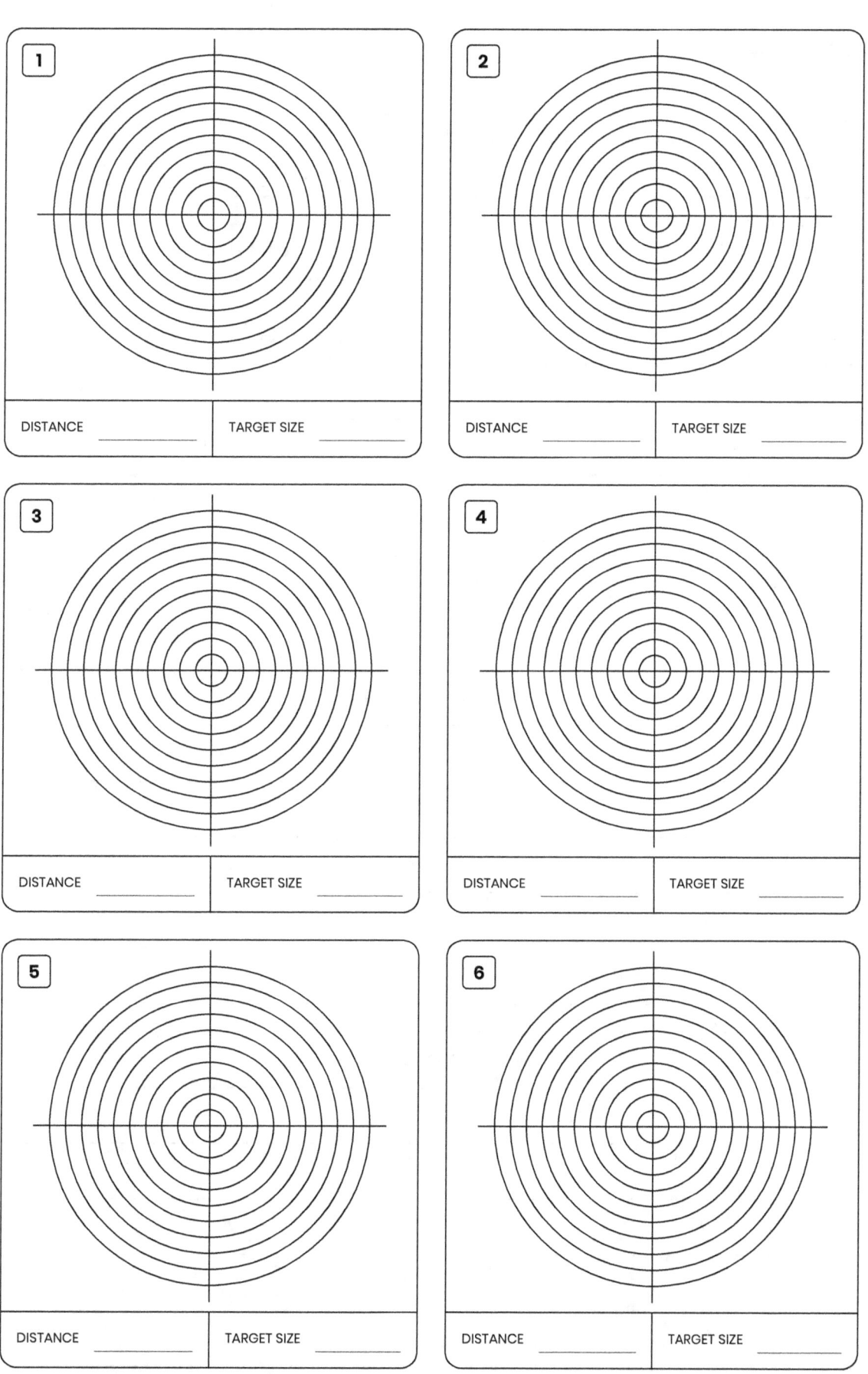

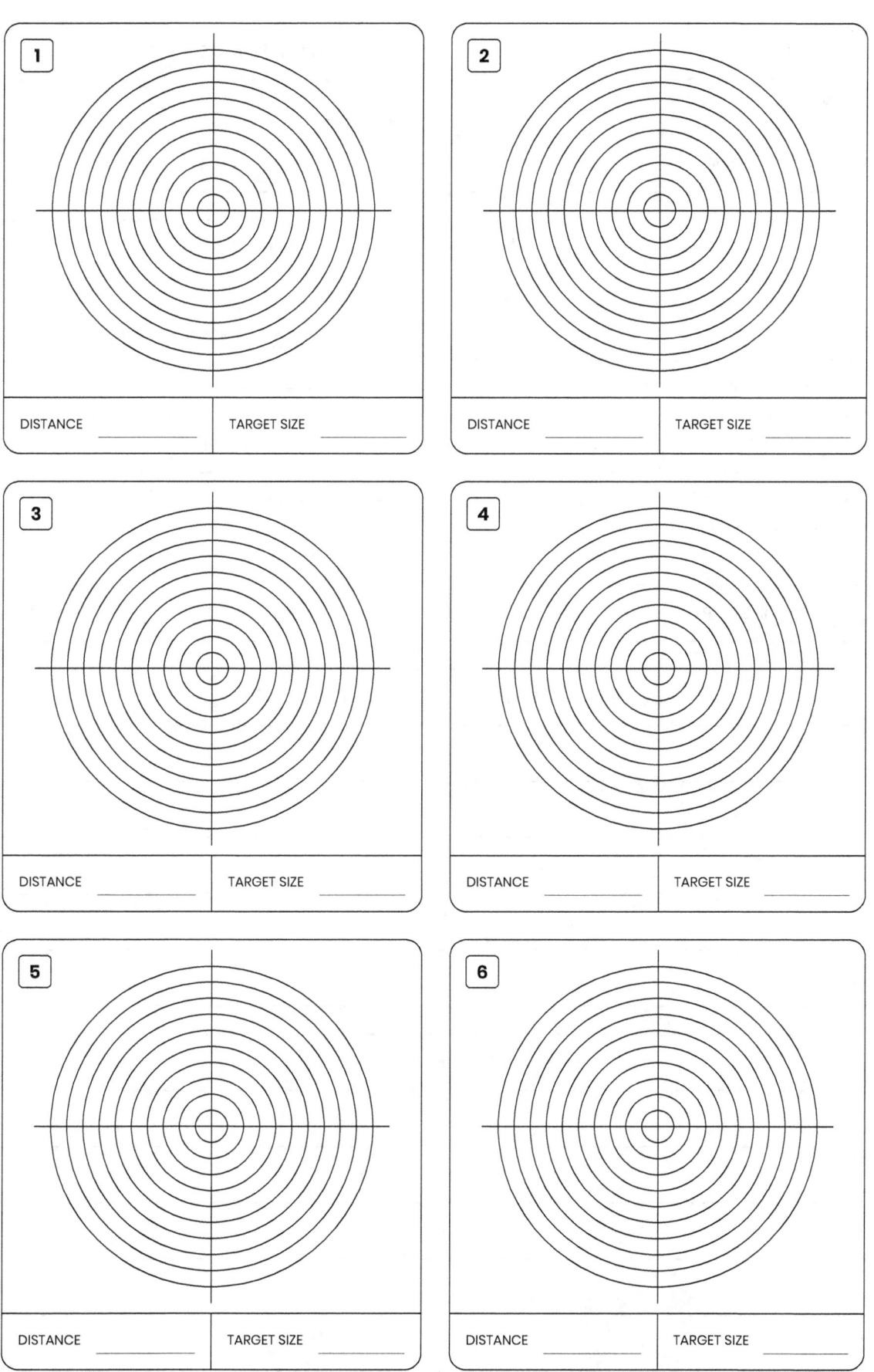

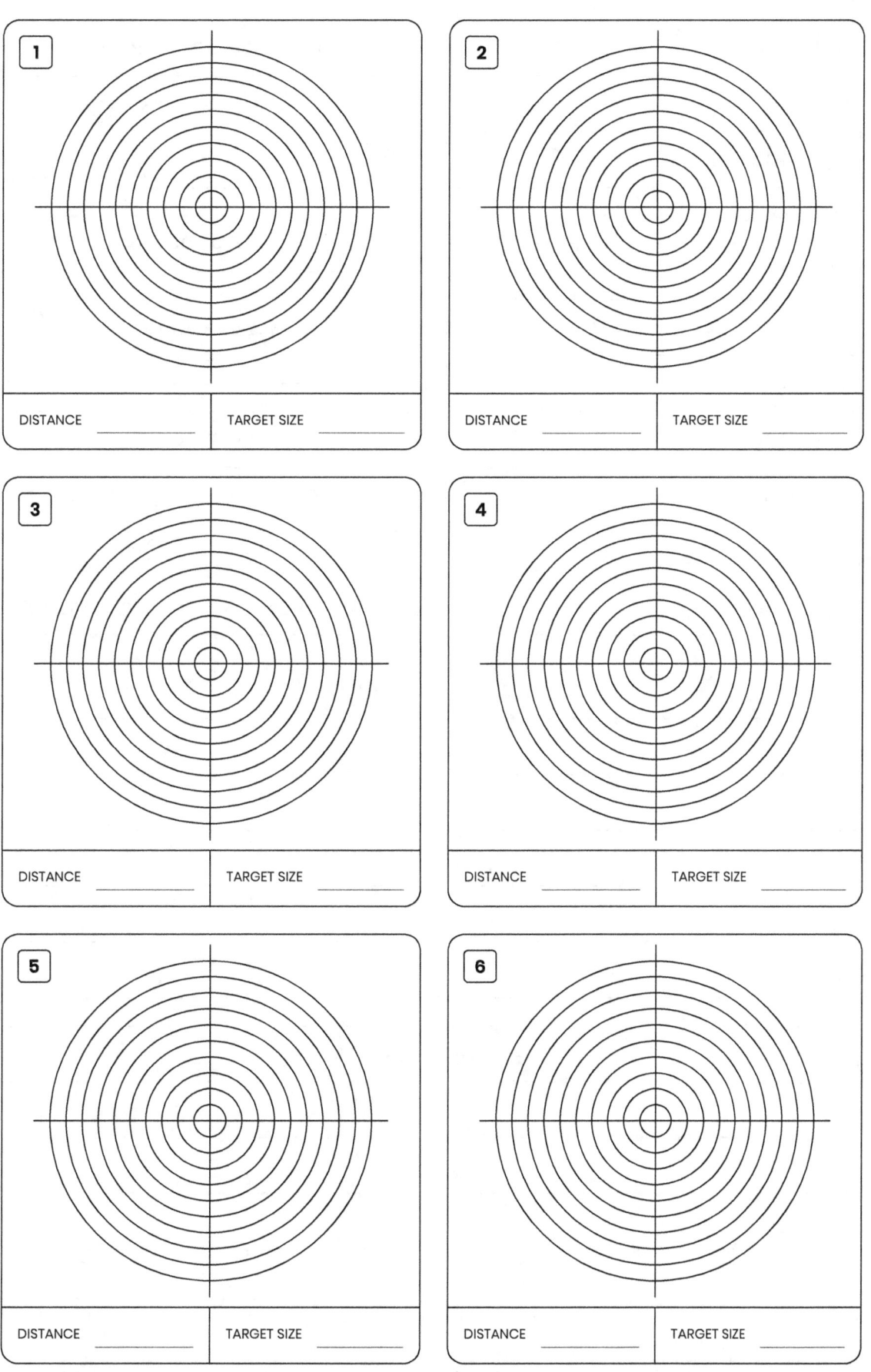

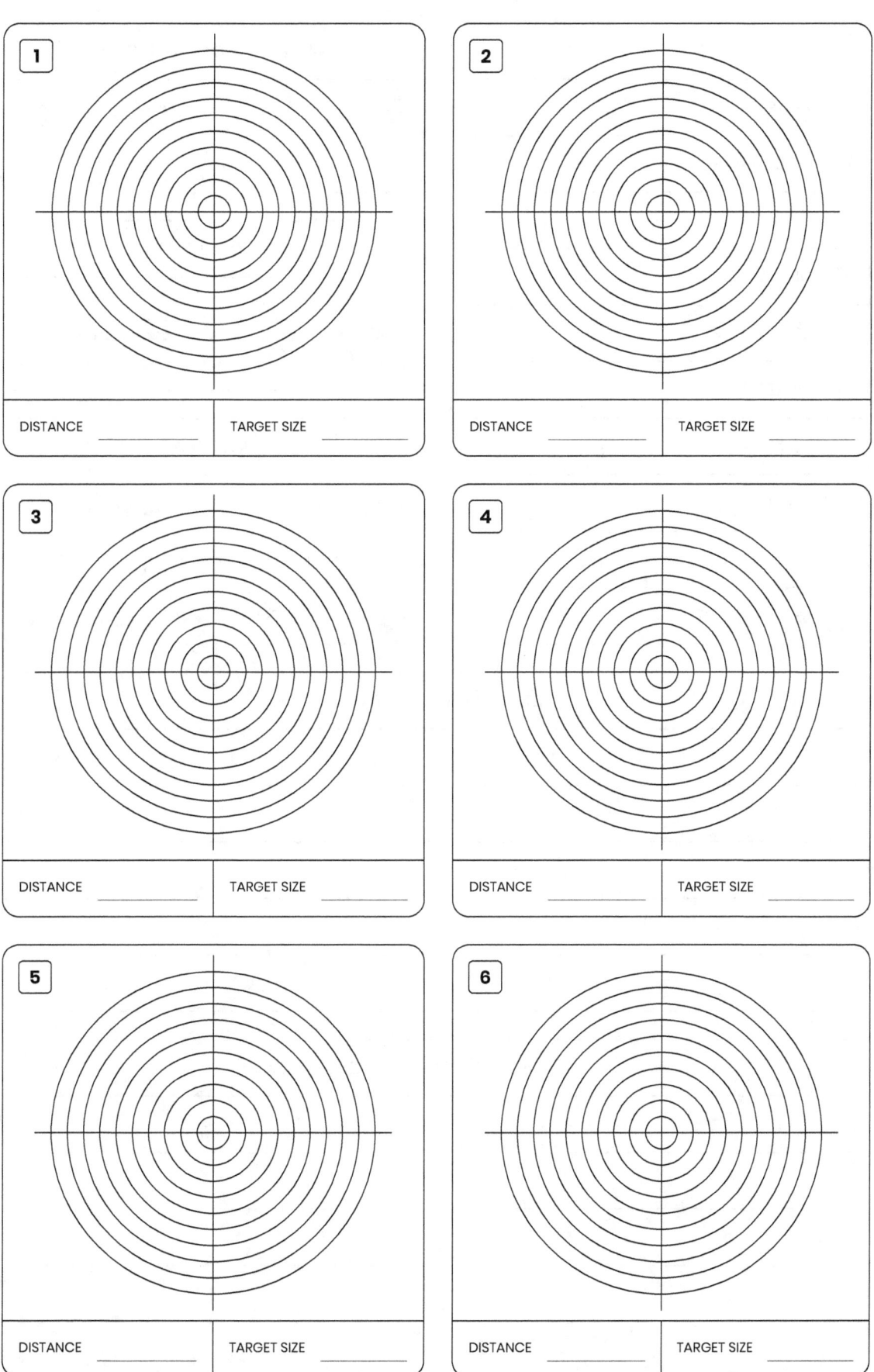

📅 DATE	
🕐 TIME	
📍 LOCATION	
👥 PARTNER	
🔫 FIREARM	
🎯 SCOPE TYPE	
🔩 BULLET TYPE	
POWDER	
PRIMER	
BRASS	
SEATING DEPTH	

WEATHER CONDITIONS

🌡 ____   ☀ ⛅ 🌧 ⛈ ❄
🚩 ____   ☐ ☐ ☐ ☐ ☐

LIGHT CONDITIONS

☀ — 1 — 2 — 3 — 4 — 5 — 🌙
BRIGHT   ○   ○   ○   ○   ○   DARK

RATING

WEAPON HANDLING		☆☆☆☆☆
HIT RATE		☆☆☆☆☆
OVERALL RESULTS		☆☆☆☆☆

SESSION HIGHLIGHTS

ADDITIONAL NOTES

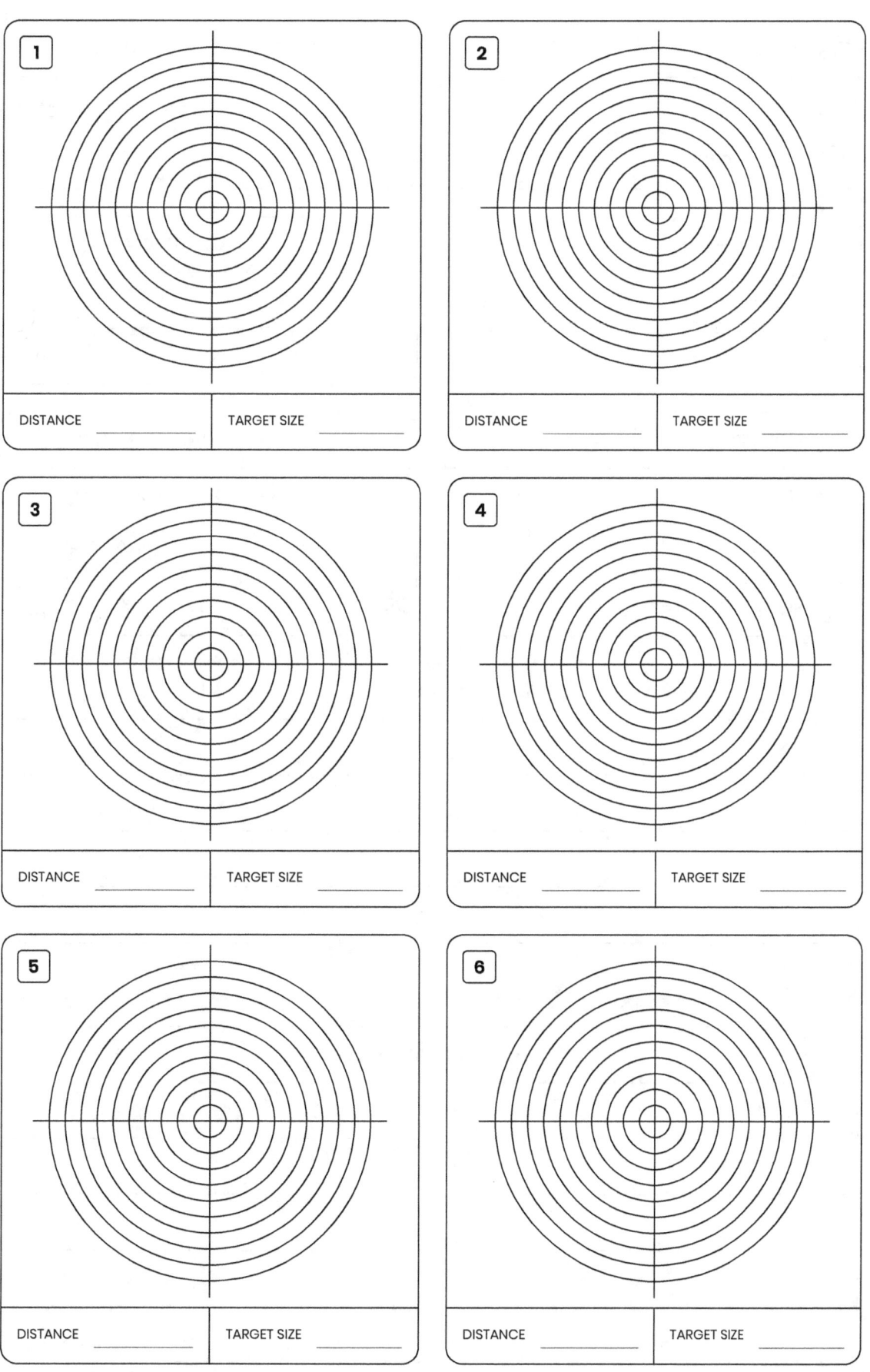

📅	**DATE**
🕐	**TIME**
📍	**LOCATION**
👥	**PARTNER**
🔫	**FIREARM**
🎯	**SCOPE TYPE**
🔹	**BULLET TYPE**
	POWDER
	PRIMER
	BRASS
	SEATING DEPTH

WEATHER CONDITIONS

🌡 ____   ☀ ☁ 🌧 ⛈ ❄
🚩 ____   ☐ ☐ ☐ ☐ ☐

LIGHT CONDITIONS

☀ — 1 — 2 — 3 — 4 — 5 — 🌙
BRIGHT ○ ○ ○ ○ ○ DARK

RATING

🔫	WEAPON HANDLING	☆☆☆☆☆
🎯	HIT RATE	☆☆☆☆☆
📋	OVERALL RESULTS	☆☆☆☆☆

SESSION HIGHLIGHTS

_____
_____
_____
_____
_____

ADDITIONAL NOTES

_____
_____
_____
_____
_____

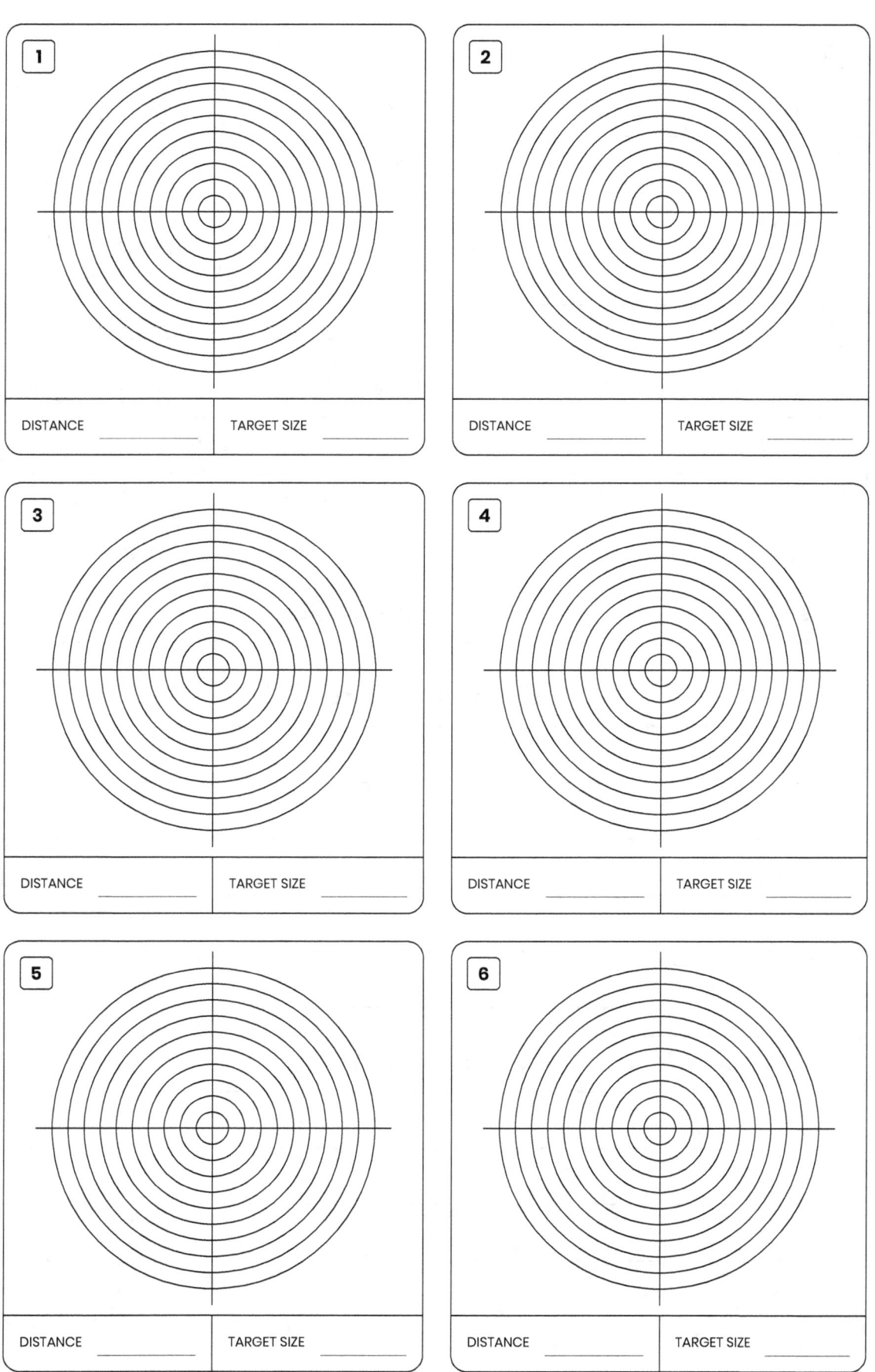

📅	**DATE**
🕐	**TIME**
📍	**LOCATION**
👥	PARTNER
🔫	FIREARM
🎯	SCOPE TYPE
🔸	BULLET TYPE
	POWDER
	PRIMER
	BRASS
	SEATING DEPTH

WEATHER CONDITIONS

🌡️ ____   ☀️ ⛅ 🌧️ ⛈️ ❄️
🚩 ____   ☐ ☐ ☐ ☐ ☐

LIGHT CONDITIONS

☀️ 1 — 2 — 3 — 4 — 5 🌙
BRIGHT ○ ○ ○ ○ ○ DARK

RATING

🔫	WEAPON HANDLING	☆☆☆☆☆
🎯	HIT RATE	☆☆☆☆☆
📋	OVERALL RESULTS	☆☆☆☆☆

SESSION HIGHLIGHTS

ADDITIONAL NOTES

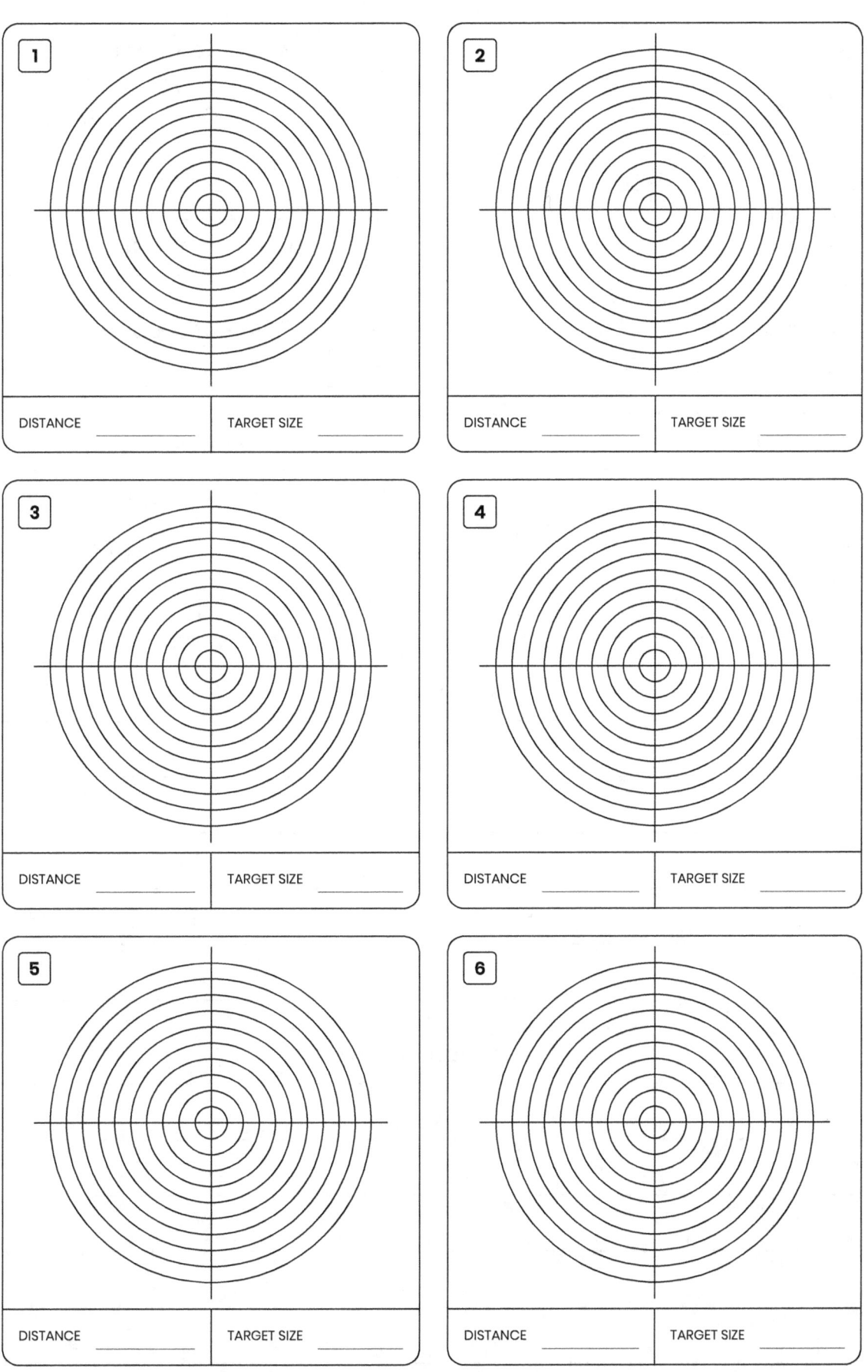

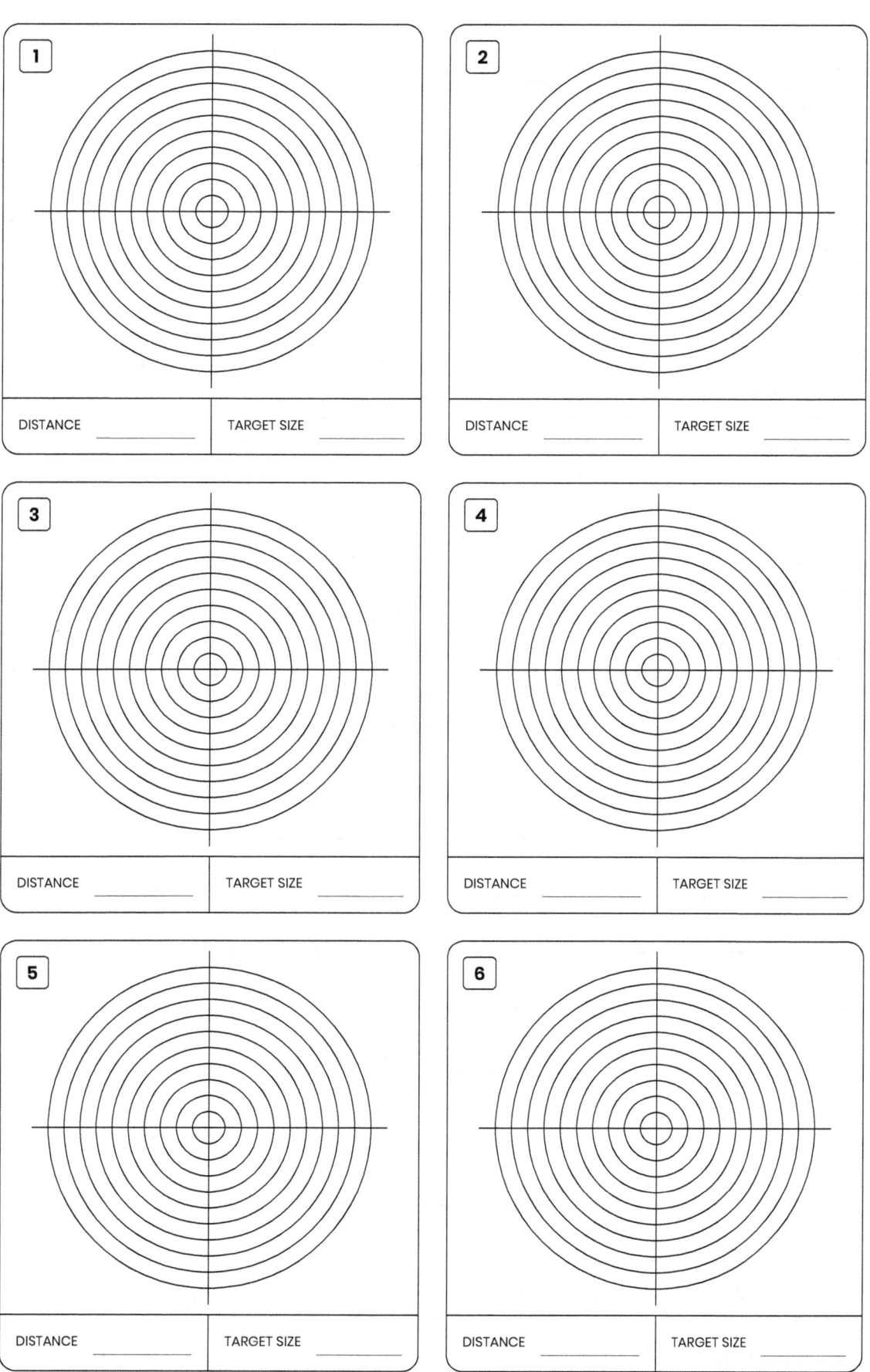

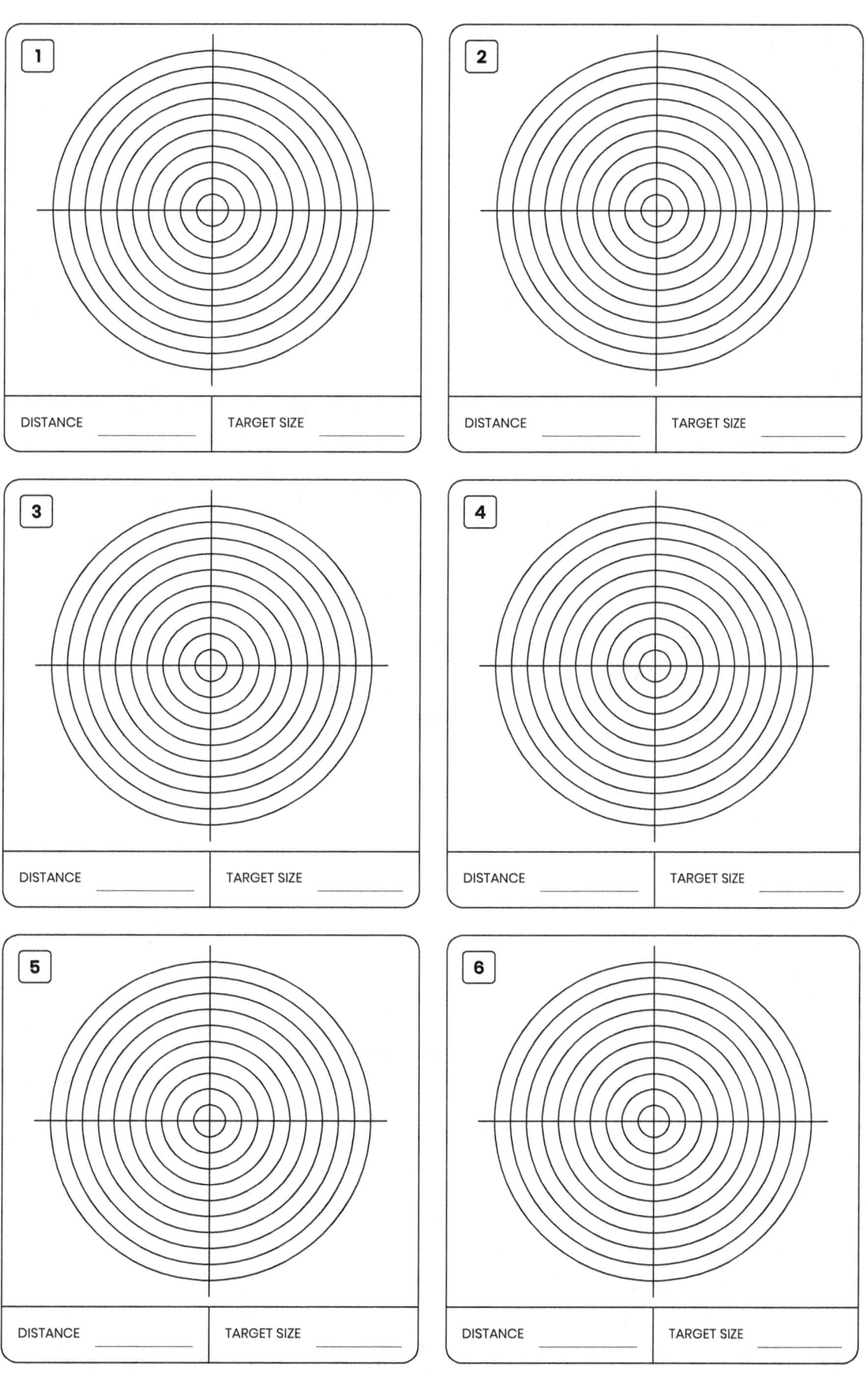

📅	**DATE**
🕐	**TIME**
📍	**LOCATION**
👥	**PARTNER**
🔫	**FIREARM**
🎯	**SCOPE TYPE**
🔹	**BULLET TYPE**
	POWDER
	PRIMER
	BRASS
	SEATING DEPTH

WEATHER CONDITIONS

🌡 ____   ☀ ⛅ 🌧 ⛈ ❄
🚩 ____   ☐ ☐ ☐ ☐ ☐

LIGHT CONDITIONS

☀ BRIGHT   1 — 2 — 3 — 4 — 5   🌙 DARK

RATING

	WEAPON HANDLING	☆☆☆☆☆
	HIT RATE	☆☆☆☆☆
	OVERALL RESULTS	☆☆☆☆☆

SESSION HIGHLIGHTS

_____
_____
_____
_____
_____

ADDITIONAL NOTES

_____
_____
_____
_____
_____

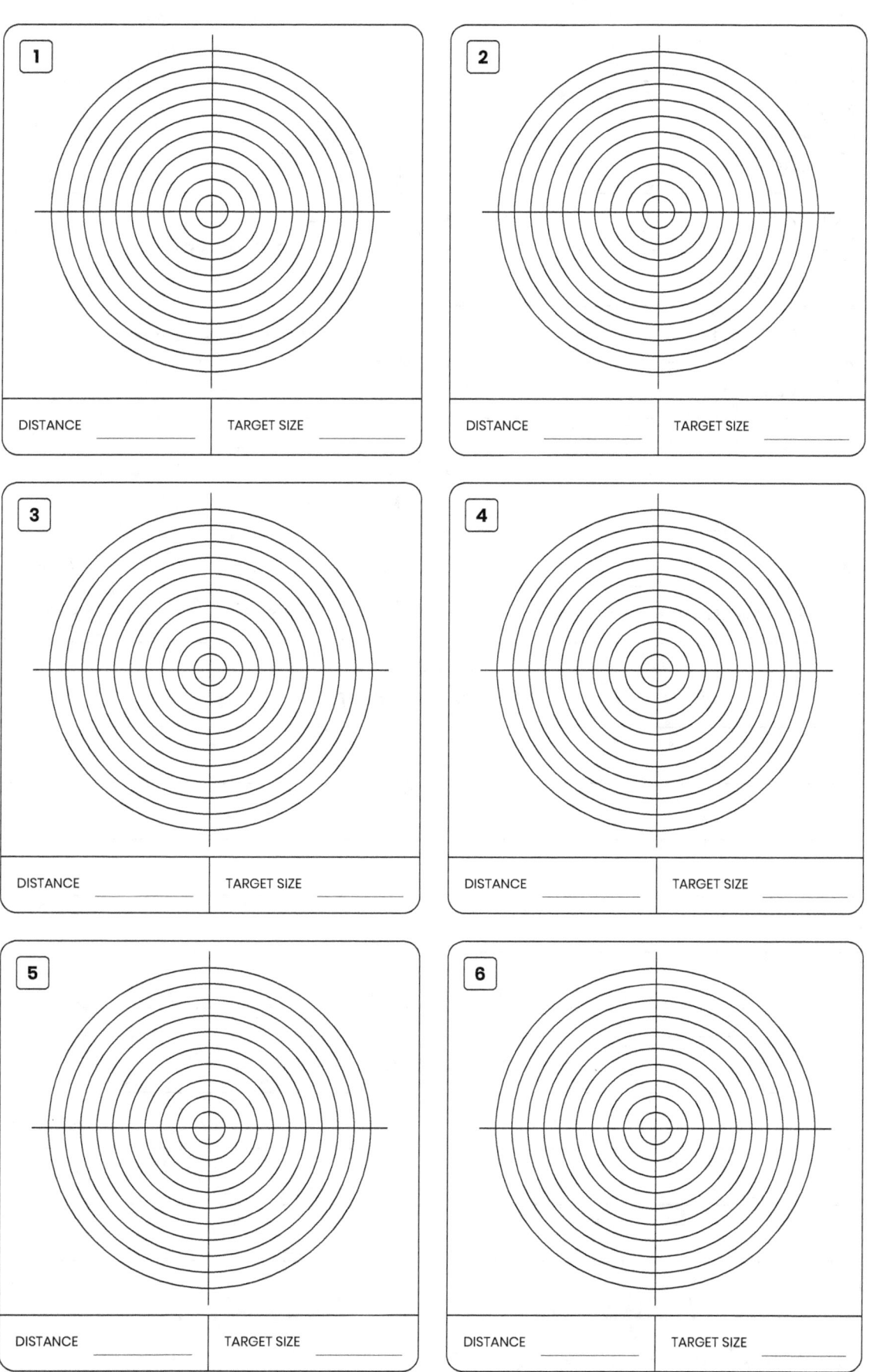

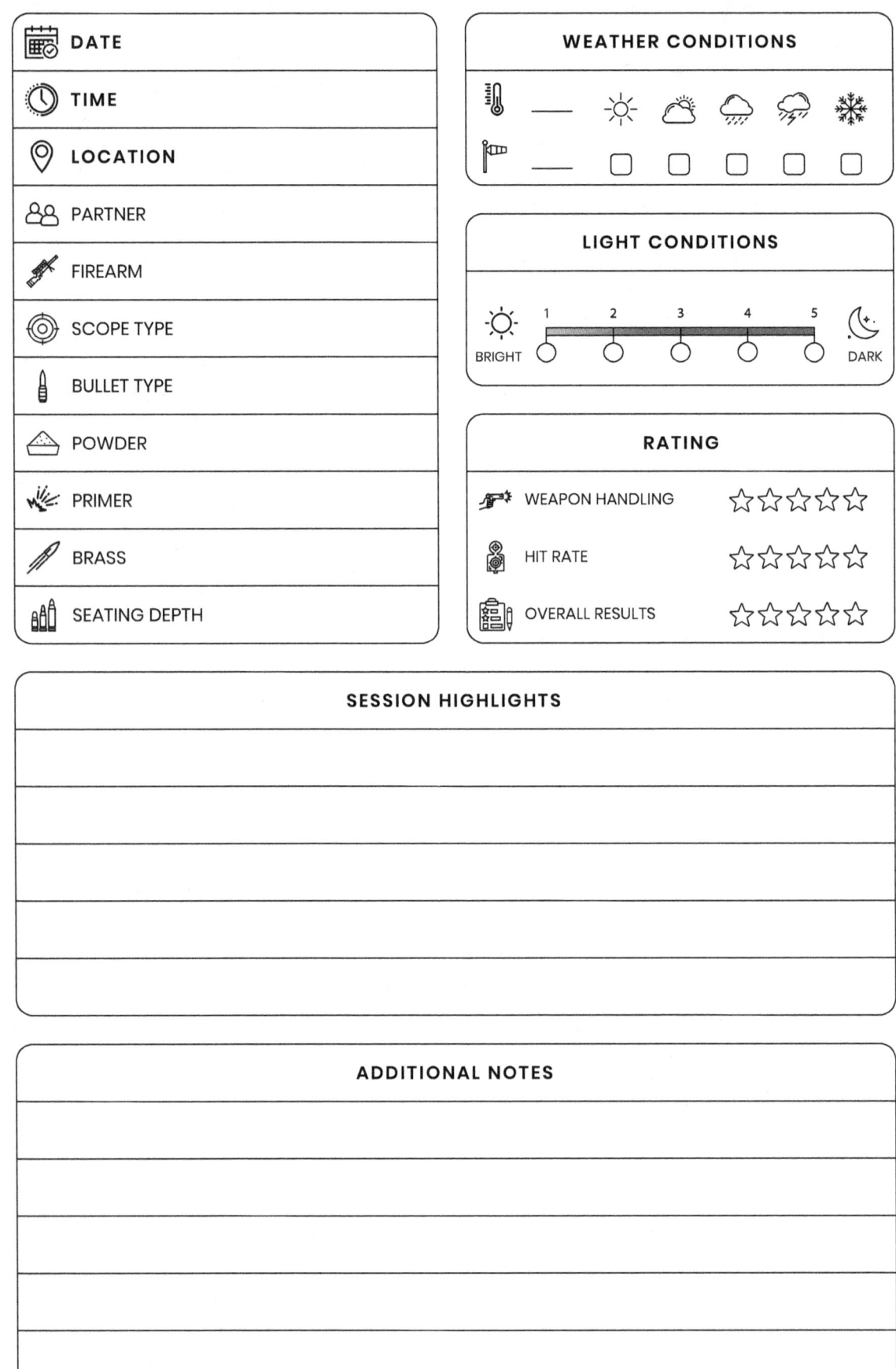

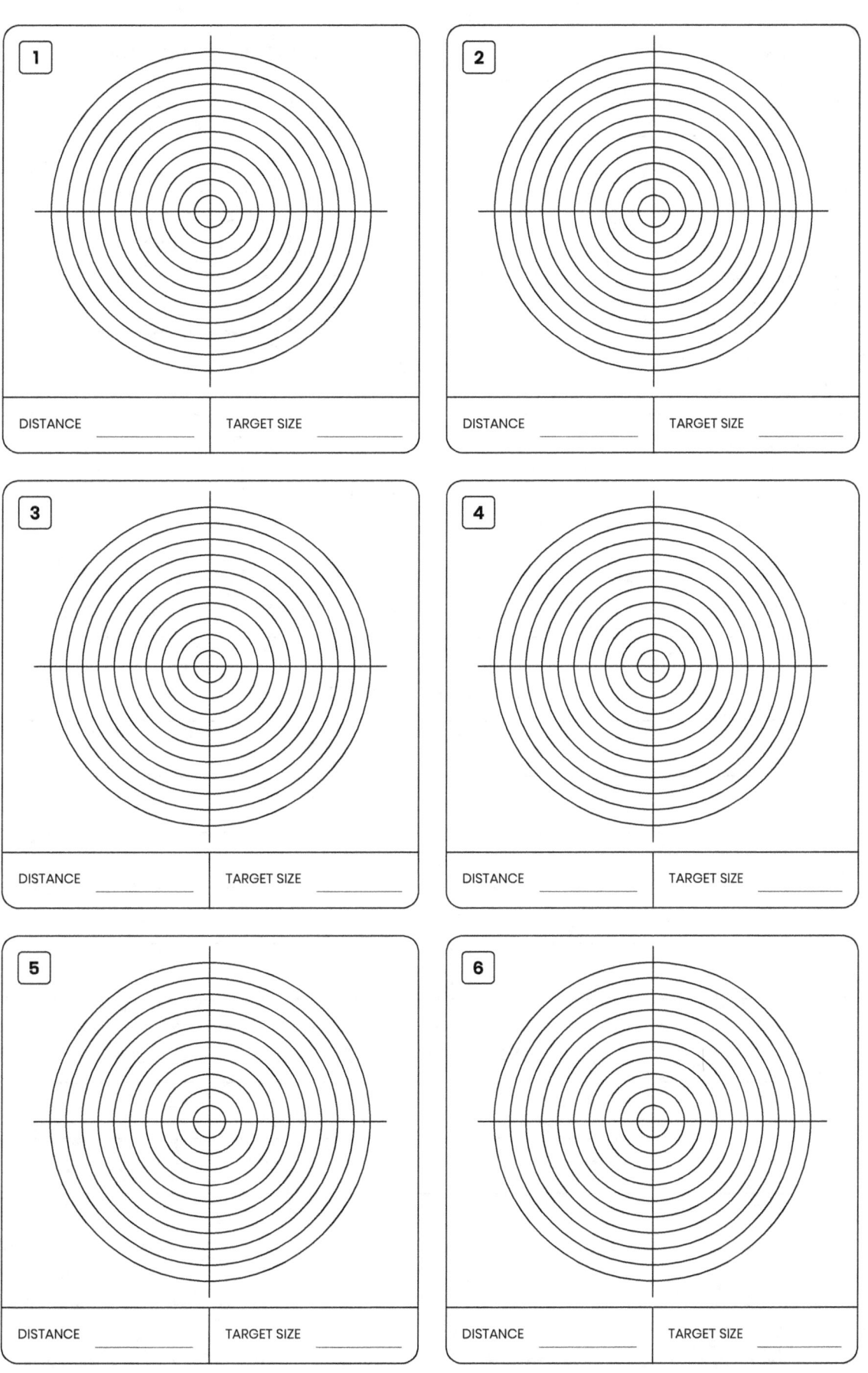

📅 DATE	
🕐 TIME	
📍 LOCATION	
👥 PARTNER	
🔫 FIREARM	
🎯 SCOPE TYPE	
🔸 BULLET TYPE	
⛰ POWDER	
💥 PRIMER	
✒ BRASS	
🔫 SEATING DEPTH	

WEATHER CONDITIONS

🌡 _____ ☀ 🌤 🌧 ⛈ ❄
🚩 _____ ☐ ☐ ☐ ☐ ☐

LIGHT CONDITIONS

☀ 1 — 2 — 3 — 4 — 5 🌙
BRIGHT ○ ○ ○ ○ ○ DARK

RATING

🔫 WEAPON HANDLING		☆☆☆☆☆
🎯 HIT RATE		☆☆☆☆☆
📋 OVERALL RESULTS		☆☆☆☆☆

SESSION HIGHLIGHTS

ADDITIONAL NOTES

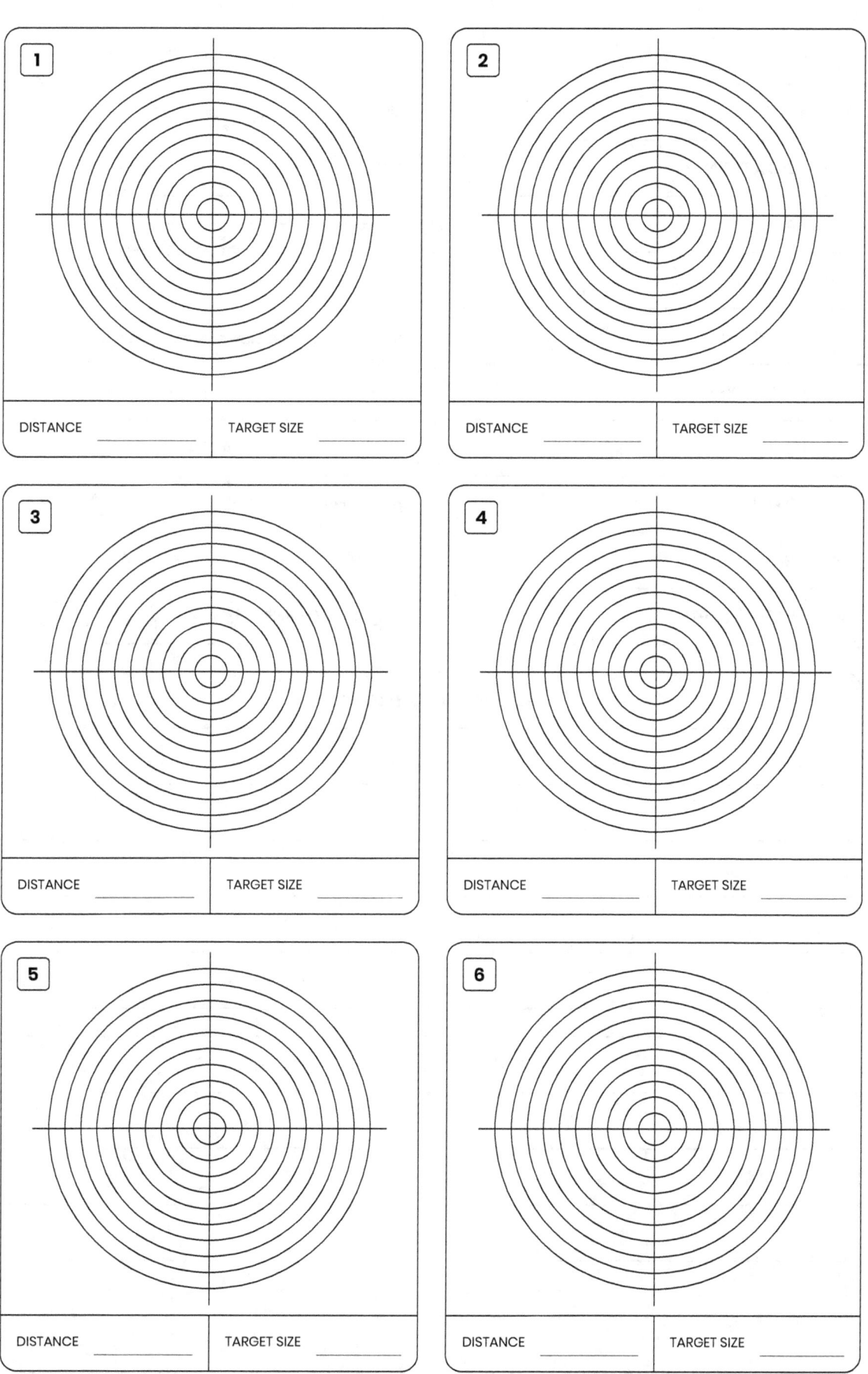

📅 **DATE**	
🕐 **TIME**	
📍 **LOCATION**	
👥 **PARTNER**	
🔫 **FIREARM**	
🎯 **SCOPE TYPE**	
🔸 **BULLET TYPE**	
⛰ **POWDER**	
💥 **PRIMER**	
🔧 **BRASS**	
🔫 **SEATING DEPTH**	

WEATHER CONDITIONS

🌡 _____ ☀ ⛅ 🌧 ⛈ ❄
🚩 _____ ☐ ☐ ☐ ☐ ☐

LIGHT CONDITIONS

☀ 1 — 2 — 3 — 4 — 5 🌙
BRIGHT   DARK

RATING

🔫	WEAPON HANDLING	☆☆☆☆☆
🎯	HIT RATE	☆☆☆☆☆
📋	OVERALL RESULTS	☆☆☆☆☆

SESSION HIGHLIGHTS

_____
_____
_____
_____
_____

ADDITIONAL NOTES

_____
_____
_____
_____
_____

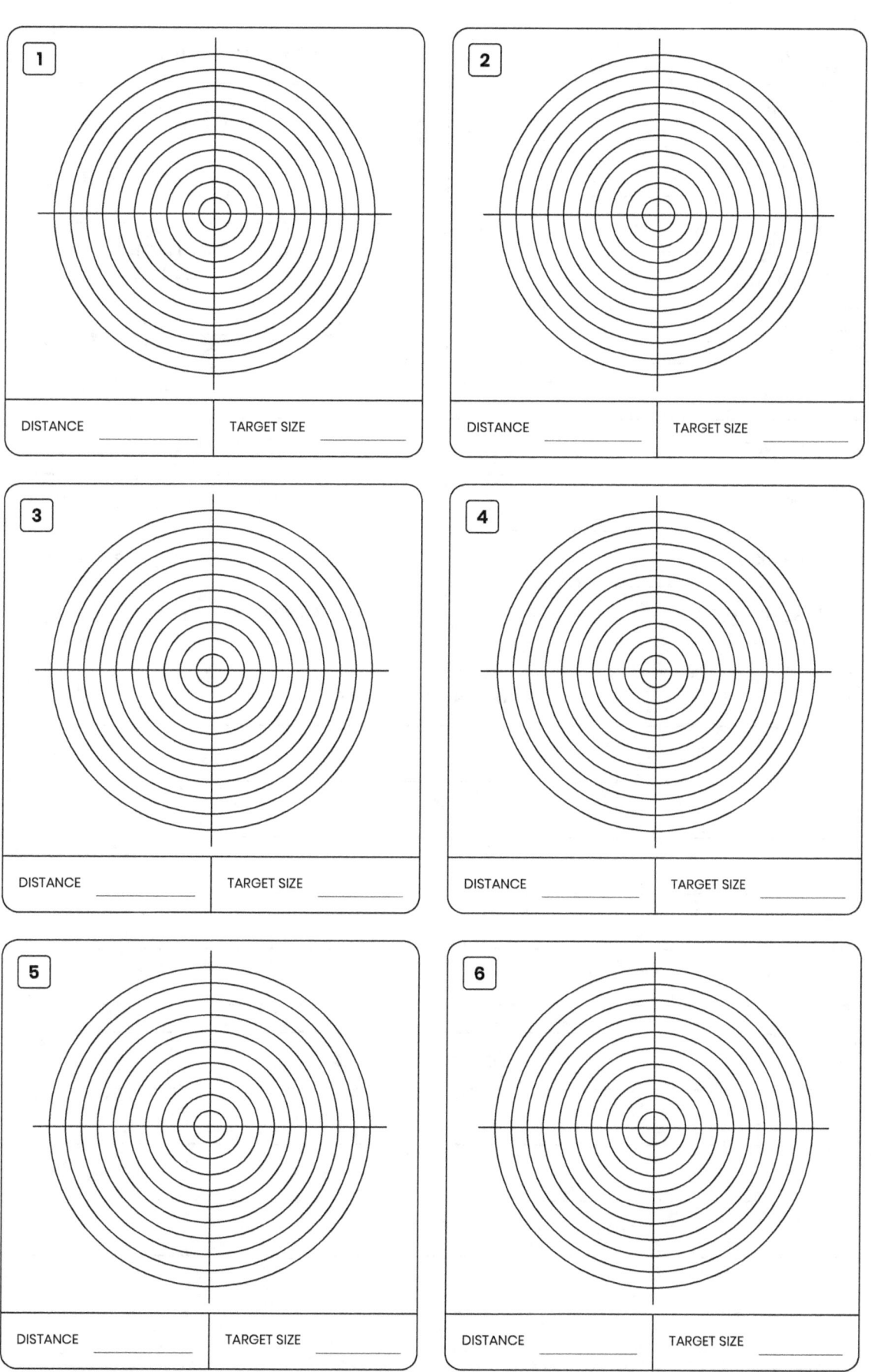

📅 DATE	
🕐 TIME	
📍 LOCATION	
👥 PARTNER	
🔫 FIREARM	
🎯 SCOPE TYPE	
🔹 BULLET TYPE	
POWDER	
PRIMER	
BRASS	
SEATING DEPTH	

WEATHER CONDITIONS

🌡 _____ ☀️ ⛅ 🌧 ⛈ ❄️

🚩 _____ ☐ ☐ ☐ ☐ ☐

LIGHT CONDITIONS

BRIGHT ☀️ 1 — 2 — 3 — 4 — 5 🌙 DARK
       ○   ○   ○   ○   ○

RATING

🔫	WEAPON HANDLING	☆☆☆☆☆
🎯	HIT RATE	☆☆☆☆☆
📋	OVERALL RESULTS	☆☆☆☆☆

SESSION HIGHLIGHTS

_____
_____
_____
_____
_____

ADDITIONAL NOTES

_____
_____
_____
_____
_____

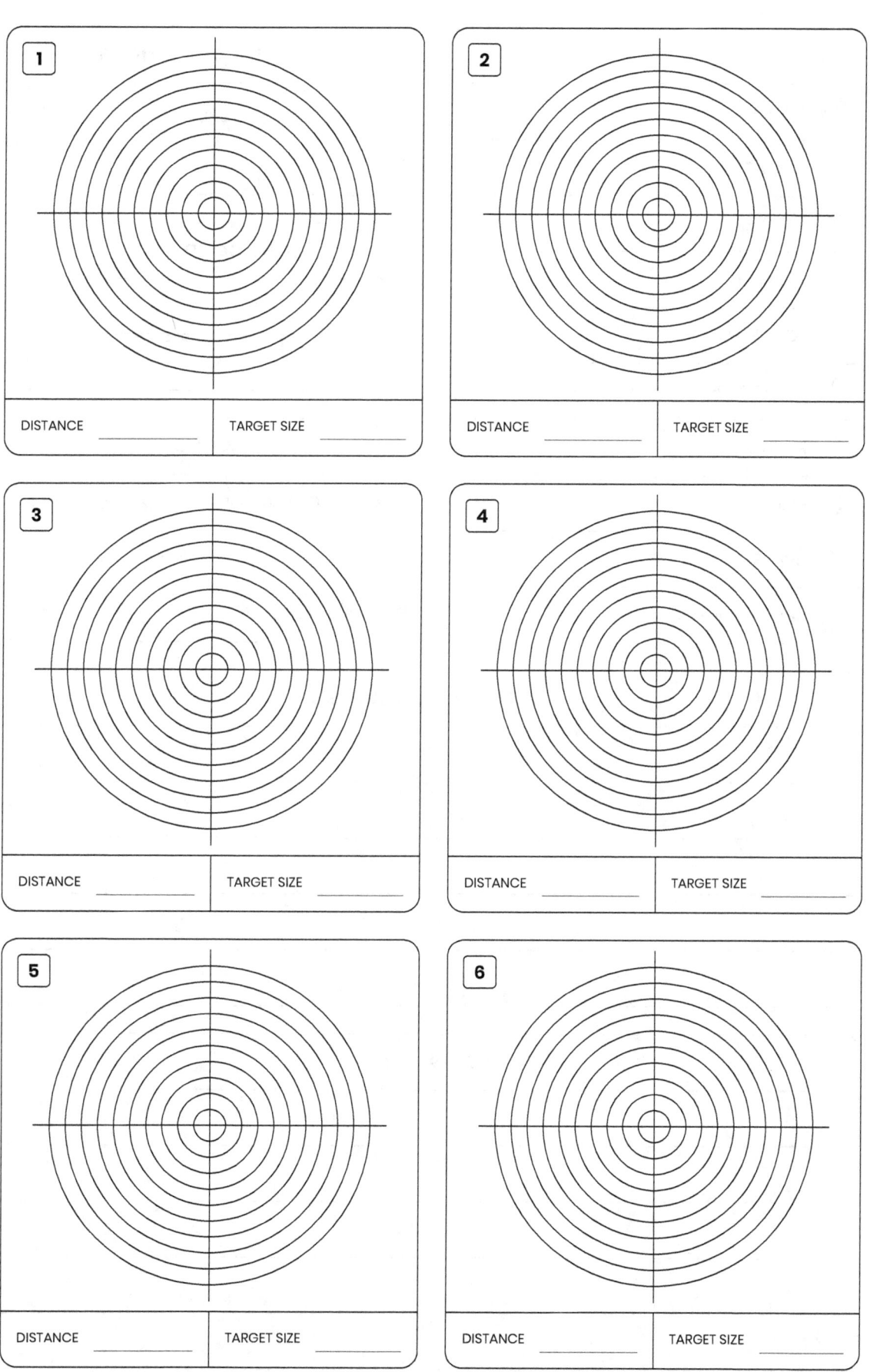

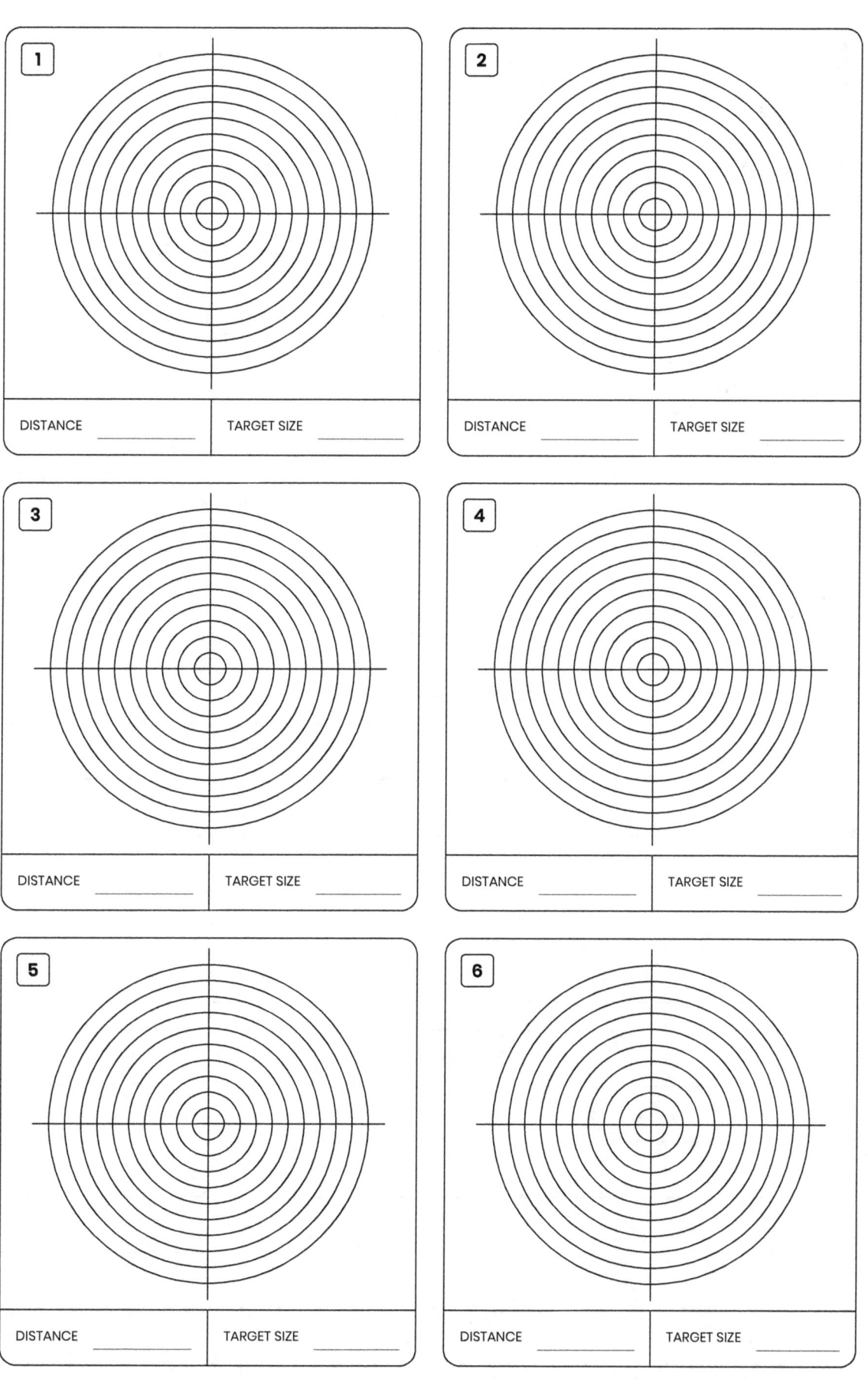

📅 DATE	
🕐 TIME	
📍 LOCATION	
👥 PARTNER	
🔫 FIREARM	
🎯 SCOPE TYPE	
🔸 BULLET TYPE	
POWDER	
PRIMER	
BRASS	
SEATING DEPTH	

WEATHER CONDITIONS

🌡 _____   ☀ ☁ 🌧 ⛈ ❄
🚩 _____   ☐ ☐ ☐ ☐ ☐

LIGHT CONDITIONS

☀ 1 — 2 — 3 — 4 — 5 🌙
BRIGHT ○   ○   ○   ○   ○ DARK

RATING

WEAPON HANDLING		☆☆☆☆☆
HIT RATE		☆☆☆☆☆
OVERALL RESULTS		☆☆☆☆☆

SESSION HIGHLIGHTS

_____
_____
_____
_____
_____

ADDITIONAL NOTES

_____
_____
_____
_____
_____

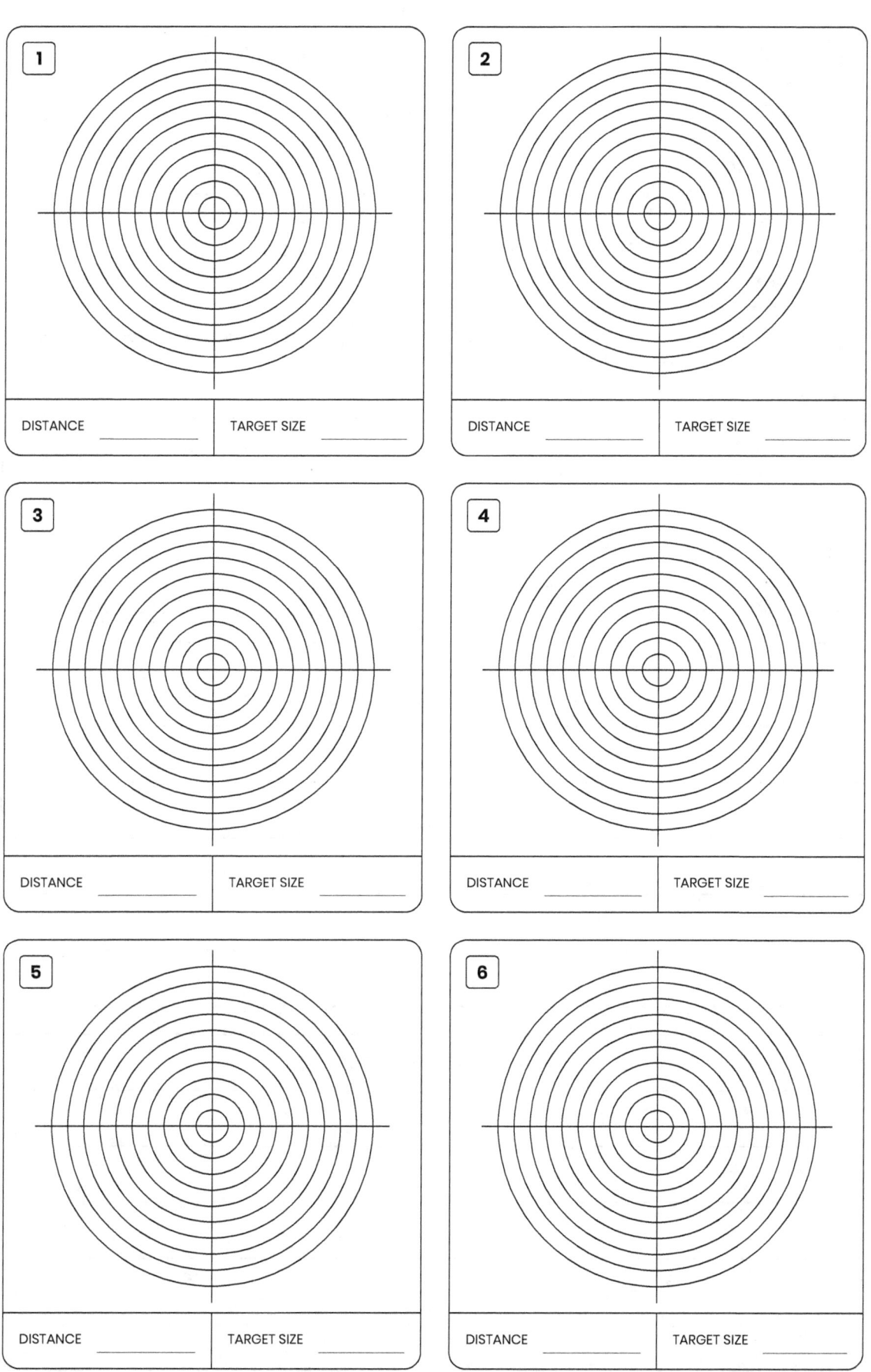

📅 DATE	
🕐 TIME	
📍 LOCATION	
👥 PARTNER	
🔫 FIREARM	
🎯 SCOPE TYPE	
🔸 BULLET TYPE	
🗂 POWDER	
💥 PRIMER	
✒ BRASS	
🔫 SEATING DEPTH	

WEATHER CONDITIONS

🌡 _____ ☀ ⛅ 🌧 ⛈ ❄
🚩 _____ ☐ ☐ ☐ ☐ ☐

LIGHT CONDITIONS

☀ BRIGHT — 1 — 2 — 3 — 4 — 5 — 🌙 DARK

RATING

🔫 WEAPON HANDLING		☆☆☆☆☆
🎯 HIT RATE		☆☆☆☆☆
📋 OVERALL RESULTS		☆☆☆☆☆

SESSION HIGHLIGHTS

ADDITIONAL NOTES

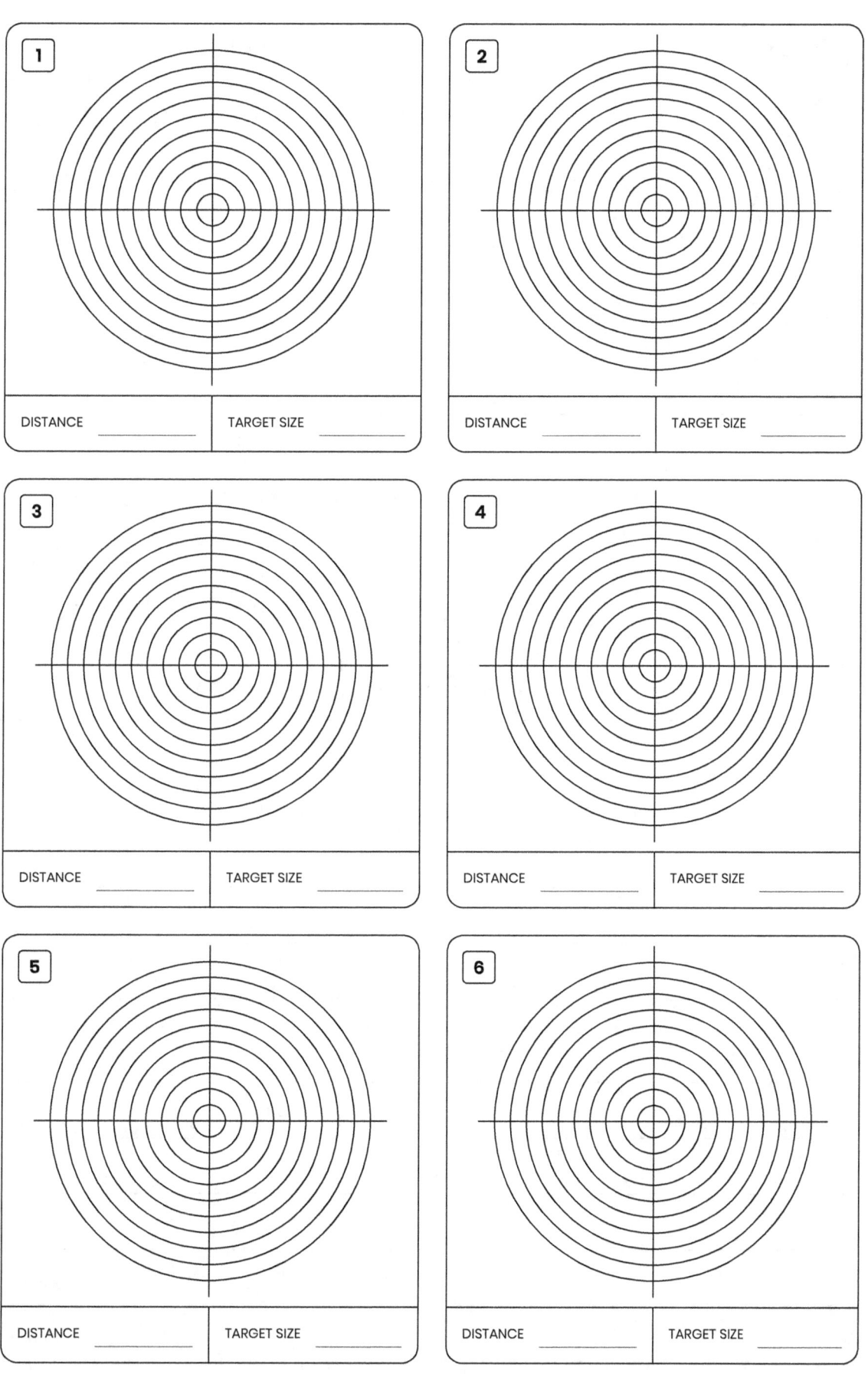

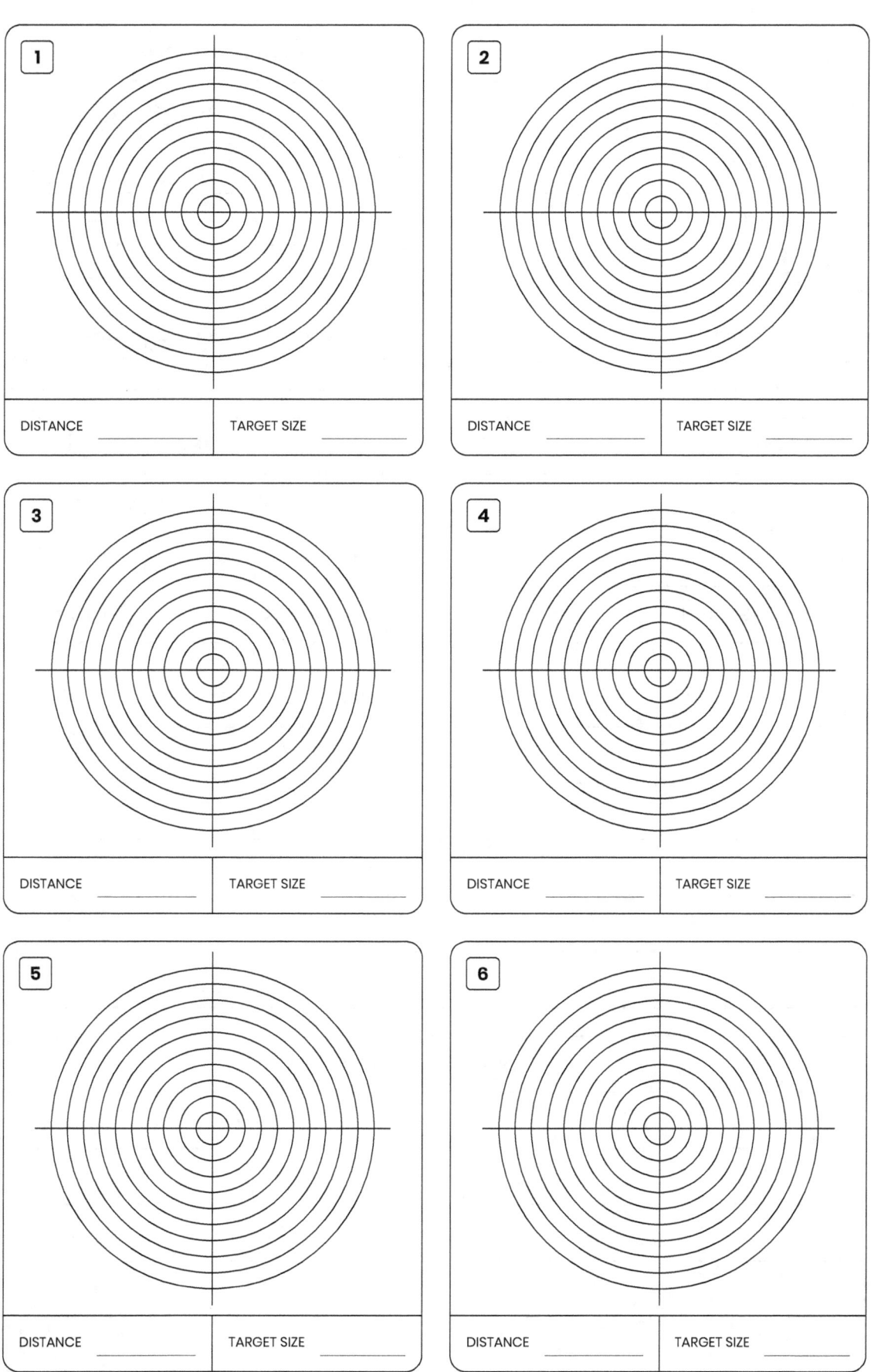

📅	**DATE**
🕐	**TIME**
📍	**LOCATION**
👥	**PARTNER**
🔫	**FIREARM**
🎯	**SCOPE TYPE**
🔸	**BULLET TYPE**
	POWDER
	PRIMER
	BRASS
	SEATING DEPTH

WEATHER CONDITIONS

☀️ ⛅ 🌧️ ⛈️ ❄️

LIGHT CONDITIONS

BRIGHT 1 — 2 — 3 — 4 — 5 DARK

RATING

- WEAPON HANDLING ☆☆☆☆☆
- HIT RATE ☆☆☆☆☆
- OVERALL RESULTS ☆☆☆☆☆

SESSION HIGHLIGHTS

ADDITIONAL NOTES

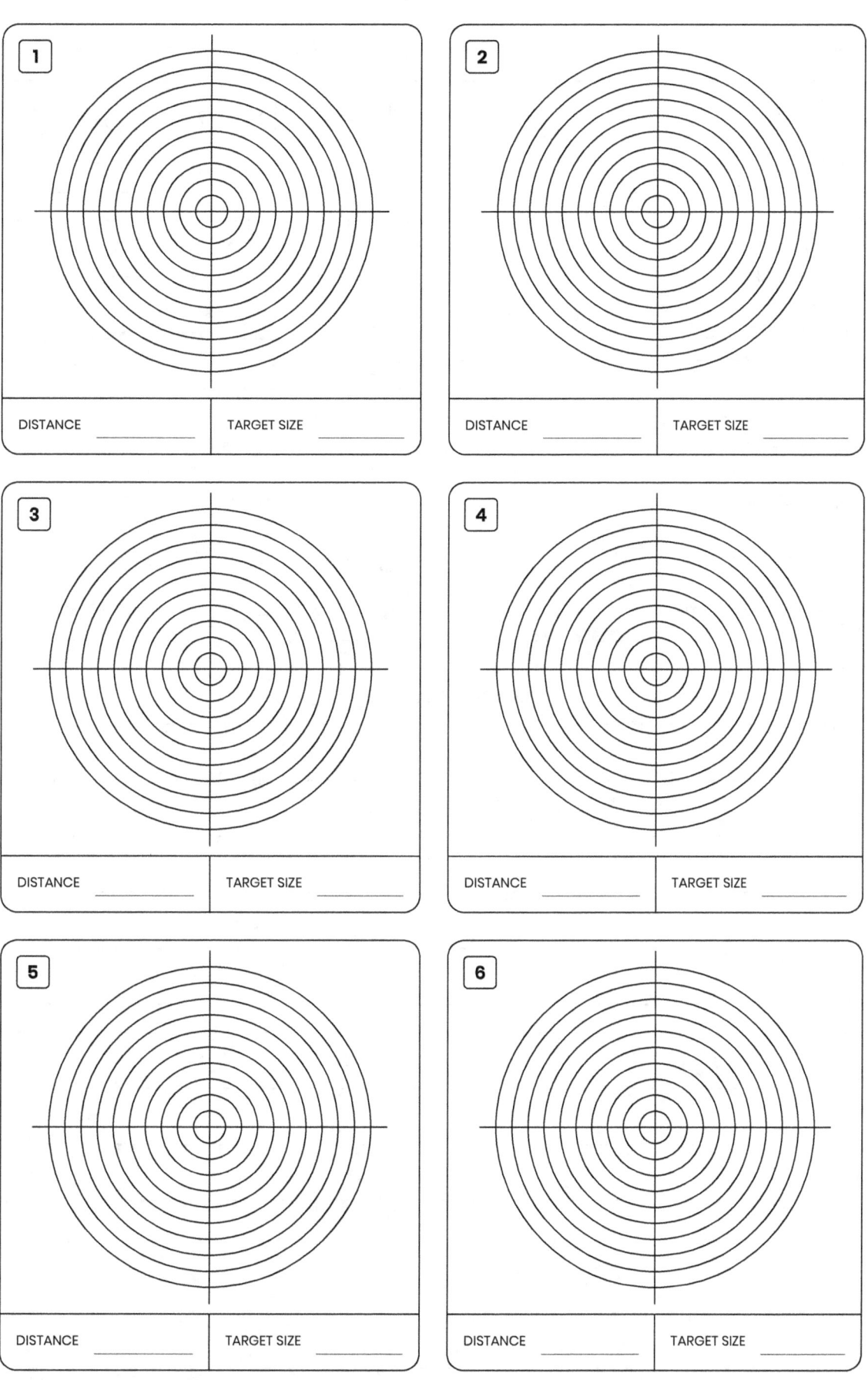

📅	DATE
🕐	TIME
📍	LOCATION
👥	PARTNER
🔫	FIREARM
🎯	SCOPE TYPE
🔸	BULLET TYPE
🥄	POWDER
💥	PRIMER
✒️	BRASS
🔫	SEATING DEPTH

WEATHER CONDITIONS

🌡️ ____   ☀️ ⛅ 🌧️ ⛈️ ❄️
🚩 ____   ☐ ☐ ☐ ☐ ☐

LIGHT CONDITIONS

☀️ 1 — 2 — 3 — 4 — 5 🌙
BRIGHT ○ ○ ○ ○ ○ DARK

RATING

🔫	WEAPON HANDLING	☆☆☆☆☆
🎯	HIT RATE	☆☆☆☆☆
📋	OVERALL RESULTS	☆☆☆☆☆

SESSION HIGHLIGHTS

ADDITIONAL NOTES

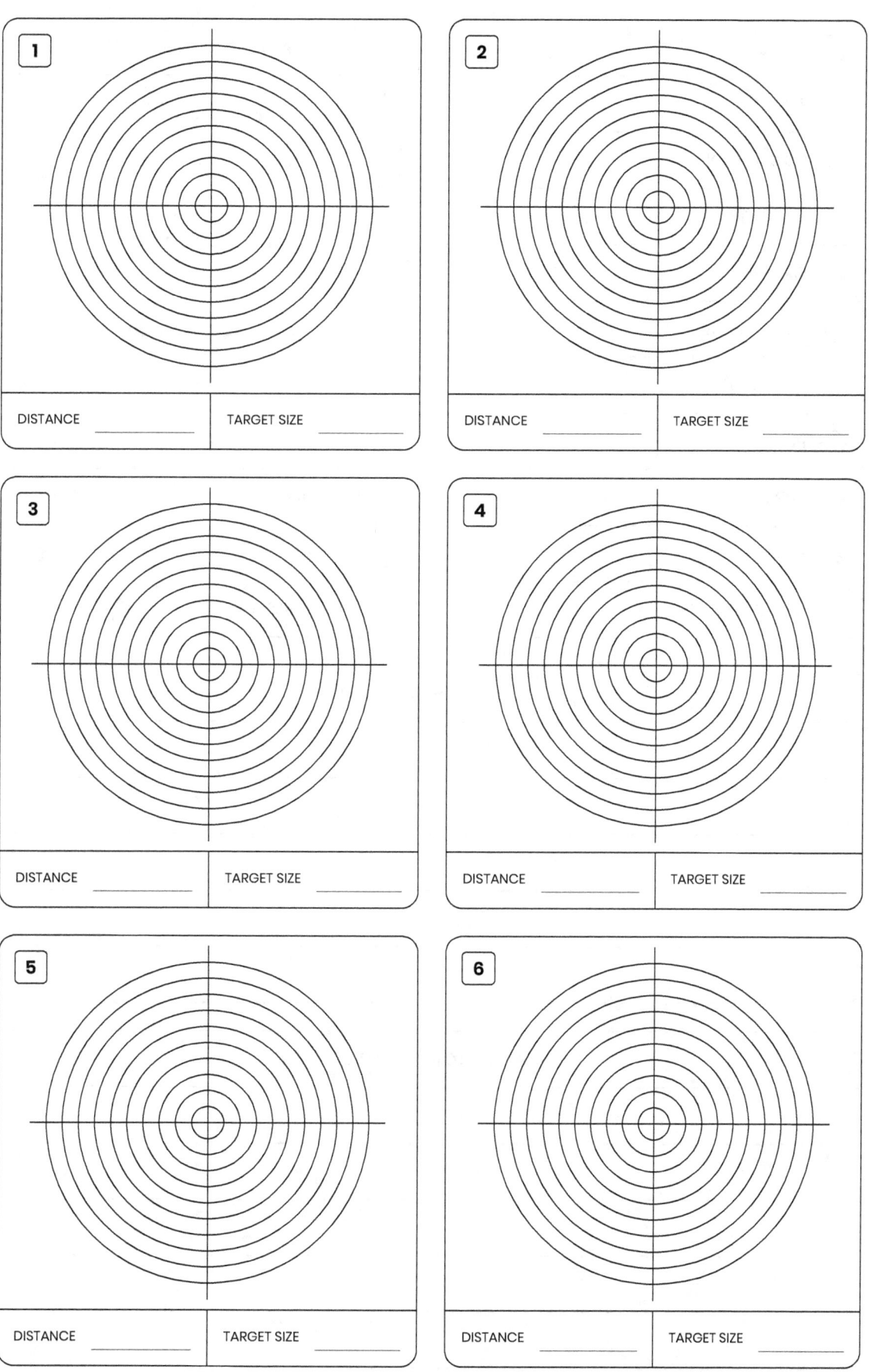

📅 DATE	
🕐 TIME	
📍 LOCATION	
👥 PARTNER	
🔫 FIREARM	
🎯 SCOPE TYPE	
🔩 BULLET TYPE	
🥄 POWDER	
💥 PRIMER	
✏️ BRASS	
🔫 SEATING DEPTH	

WEATHER CONDITIONS

🌡️ _____   ☀️ ⛅ 🌧️ ⛈️ ❄️
🚩 _____   ☐ ☐ ☐ ☐ ☐

LIGHT CONDITIONS

☀️ BRIGHT — 1 — 2 — 3 — 4 — 5 — 🌙 DARK

RATING

🔫 WEAPON HANDLING		☆☆☆☆☆
🎯 HIT RATE		☆☆☆☆☆
📋 OVERALL RESULTS		☆☆☆☆☆

SESSION HIGHLIGHTS

_____
_____
_____
_____
_____

ADDITIONAL NOTES

_____
_____
_____
_____
_____

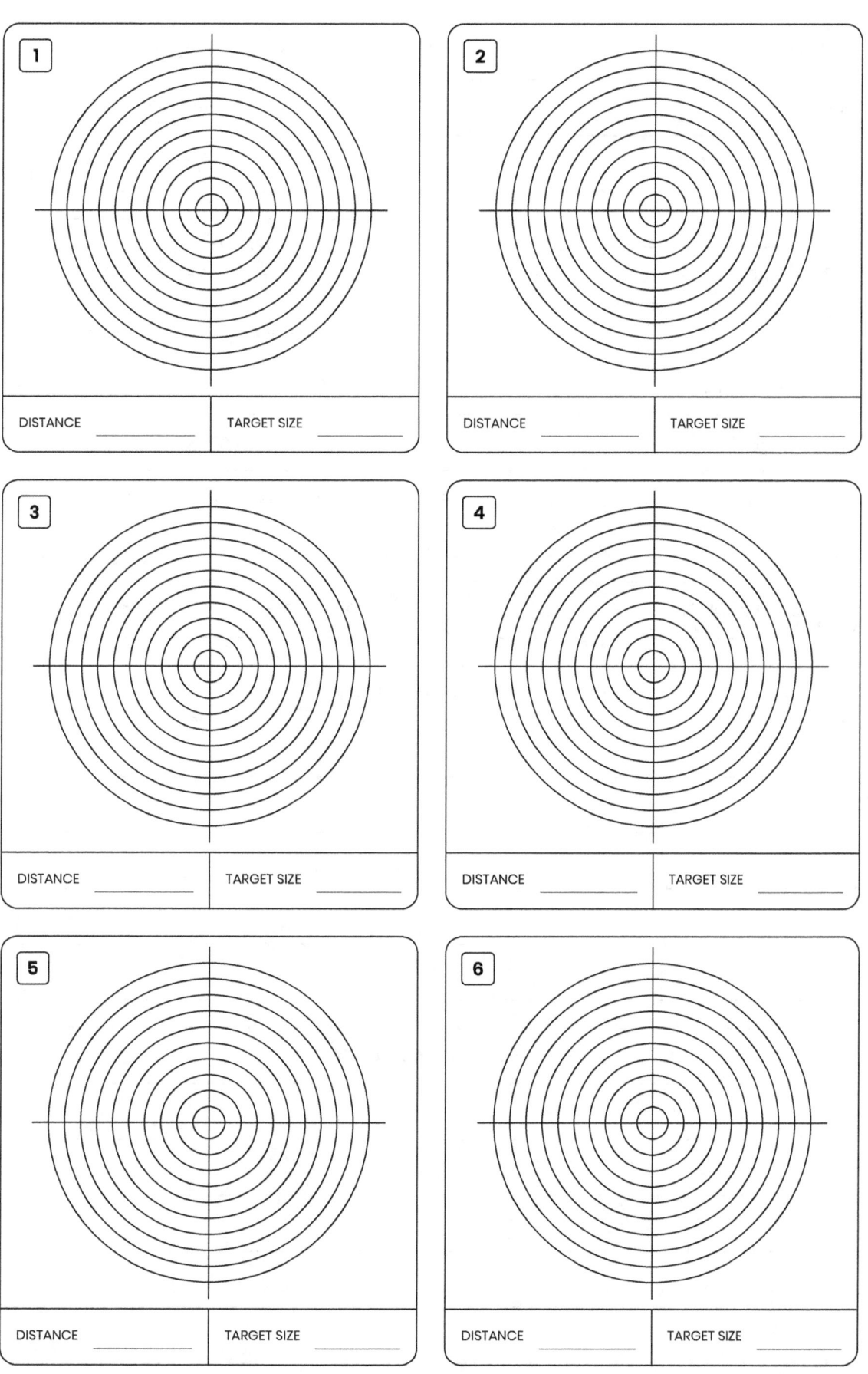

📅 DATE	
🕐 TIME	
📍 LOCATION	
👥 PARTNER	
🔫 FIREARM	
🎯 SCOPE TYPE	
🔸 BULLET TYPE	
POWDER	
PRIMER	
BRASS	
SEATING DEPTH	

WEATHER CONDITIONS

🌡 _____ ☀ ⛅ 🌧 ⛈ ❄
🚩 _____ ☐ ☐ ☐ ☐ ☐

LIGHT CONDITIONS

☀ 1 — 2 — 3 — 4 — 5 ☾
BRIGHT ○ ○ ○ ○ ○ DARK

RATING

🔫	WEAPON HANDLING	☆☆☆☆☆
🎯	HIT RATE	☆☆☆☆☆
📋	OVERALL RESULTS	☆☆☆☆☆

SESSION HIGHLIGHTS

ADDITIONAL NOTES

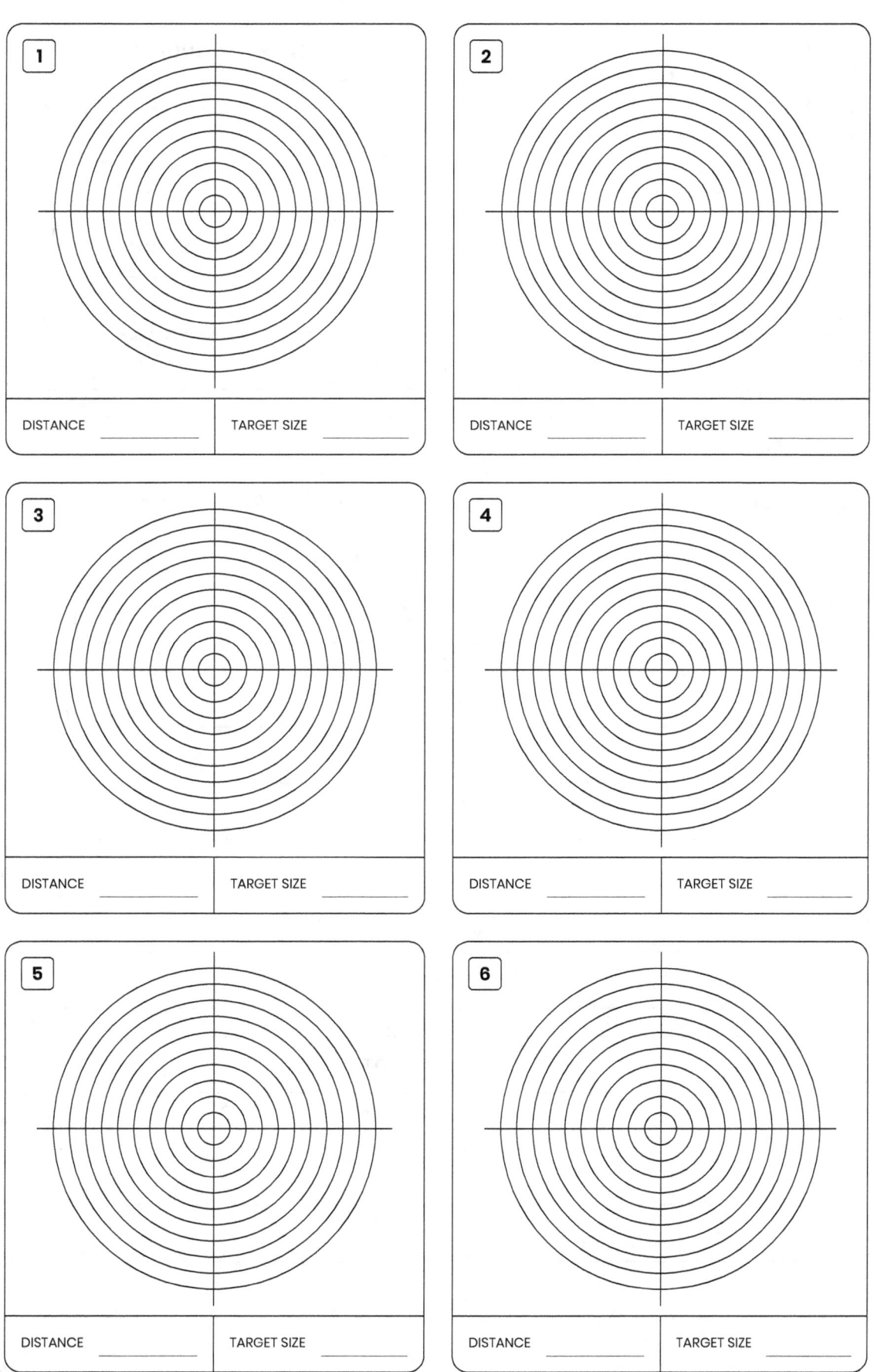

📅	**DATE**
🕐	**TIME**
📍	**LOCATION**
👥	**PARTNER**
🔫	**FIREARM**
🎯	**SCOPE TYPE**
🔸	**BULLET TYPE**
	POWDER
	PRIMER
	BRASS
	SEATING DEPTH

WEATHER CONDITIONS

🌡️ _____   ☀️ ⛅ 🌧️ ⛈️ ❄️

🚩 _____   ☐   ☐   ☐   ☐   ☐

LIGHT CONDITIONS

☀️ 1   2   3   4   5 🌙
BRIGHT ○   ○   ○   ○   ○ DARK

RATING

🔫	WEAPON HANDLING	☆☆☆☆☆
🎯	HIT RATE	☆☆☆☆☆
📋	OVERALL RESULTS	☆☆☆☆☆

SESSION HIGHLIGHTS

ADDITIONAL NOTES

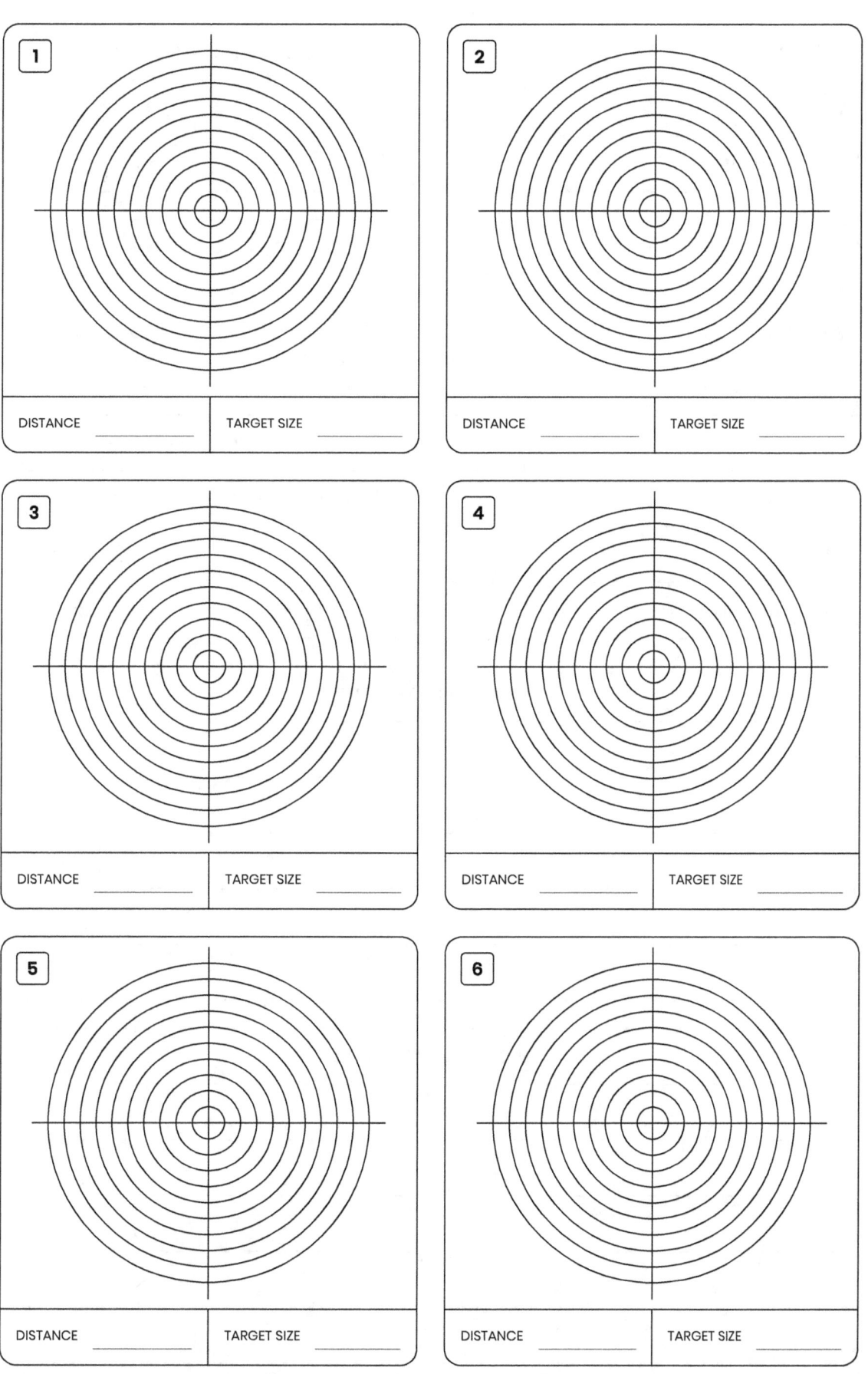

📅	**DATE**
🕐	**TIME**
📍	**LOCATION**
👥	**PARTNER**
🔫	**FIREARM**
◎	**SCOPE TYPE**
🔸	**BULLET TYPE**
	POWDER
	PRIMER
	BRASS
	SEATING DEPTH

WEATHER CONDITIONS

🌡 ____ ☀️ ⛅ 🌧 ⛈ ❄️
🚩 ____ ☐ ☐ ☐ ☐ ☐

LIGHT CONDITIONS

☀️ 1   2   3   4   5 🌙
BRIGHT ○ ○ ○ ○ ○ DARK

RATING

🔫	WEAPON HANDLING	☆☆☆☆☆
🎯	HIT RATE	☆☆☆☆☆
📋	OVERALL RESULTS	☆☆☆☆☆

SESSION HIGHLIGHTS

_____
_____
_____
_____
_____

ADDITIONAL NOTES

_____
_____
_____
_____
_____

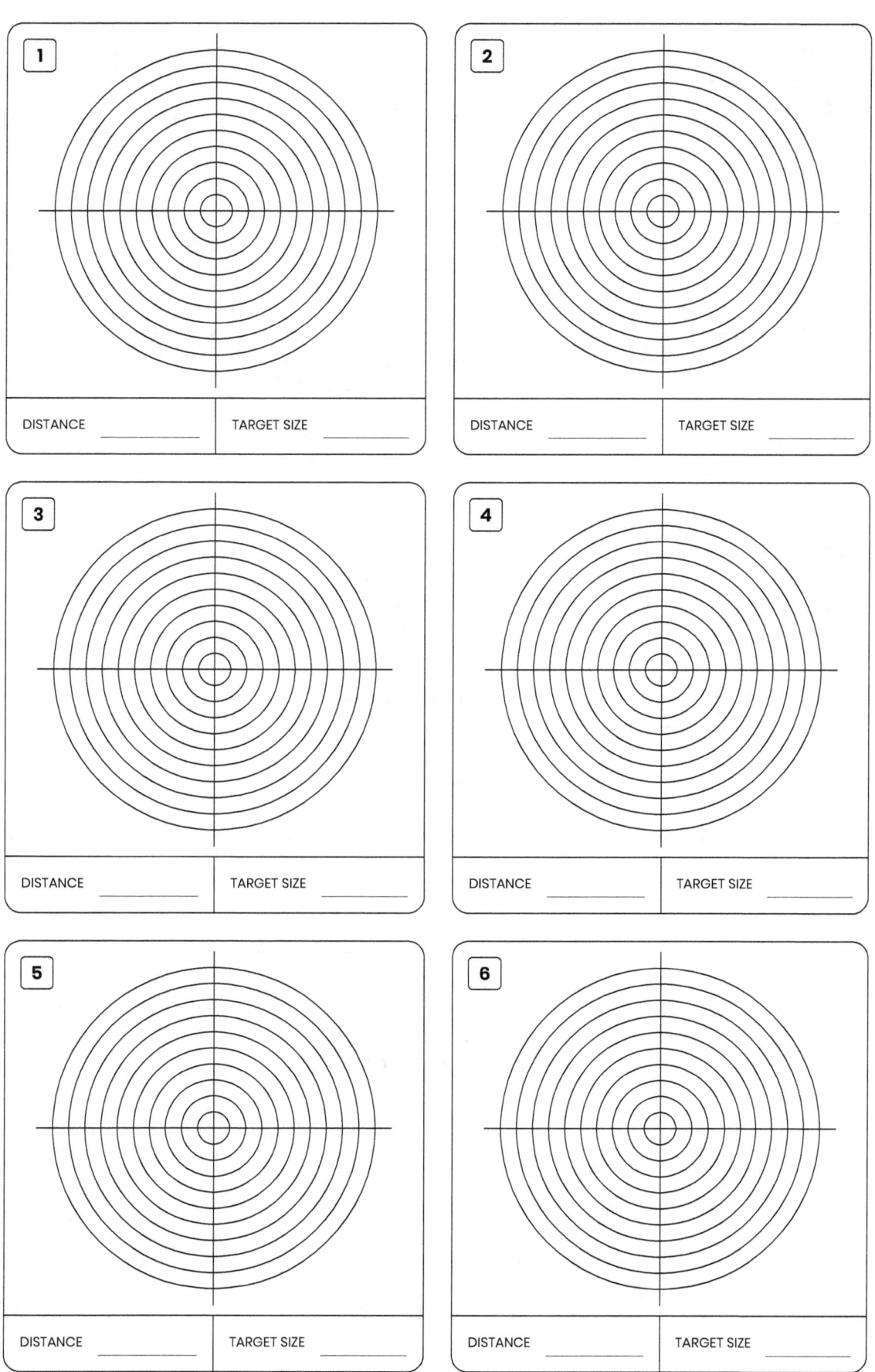

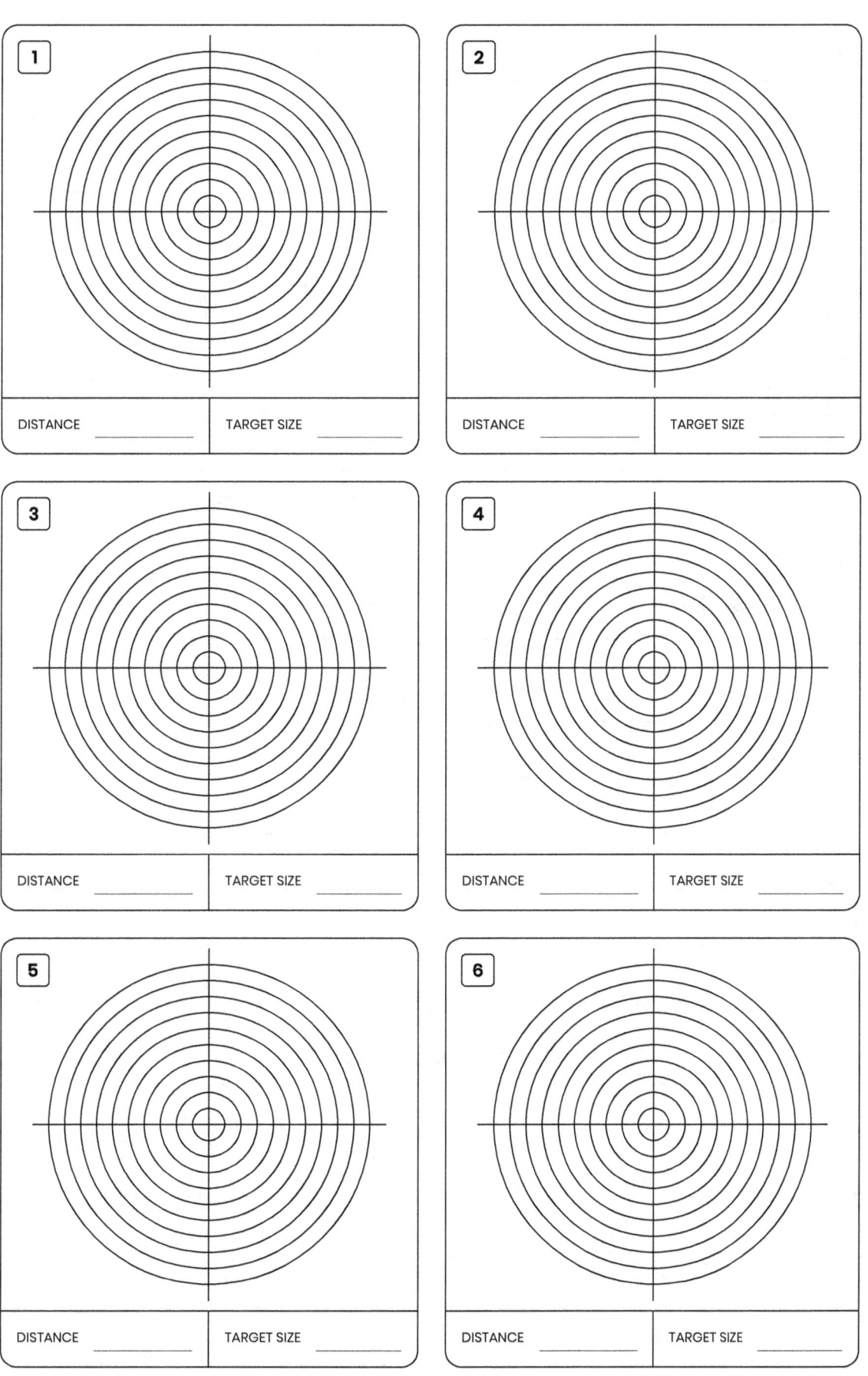

📅 DATE	
🕐 TIME	
📍 LOCATION	
👥 PARTNER	
🔫 FIREARM	
🎯 SCOPE TYPE	
🔸 BULLET TYPE	
POWDER	
PRIMER	
BRASS	
SEATING DEPTH	

WEATHER CONDITIONS

🌡 _____   ☀️ ⛅ 🌧 ⛈ ❄️
🚩 _____   ☐ ☐ ☐ ☐ ☐

LIGHT CONDITIONS

☀️ 1   2   3   4   5 🌙
BRIGHT ○ ○ ○ ○ ○ DARK

RATING

🔫	WEAPON HANDLING	☆☆☆☆☆
🎯	HIT RATE	☆☆☆☆☆
📋	OVERALL RESULTS	☆☆☆☆☆

SESSION HIGHLIGHTS

_____
_____
_____
_____
_____

ADDITIONAL NOTES

_____
_____
_____
_____
_____

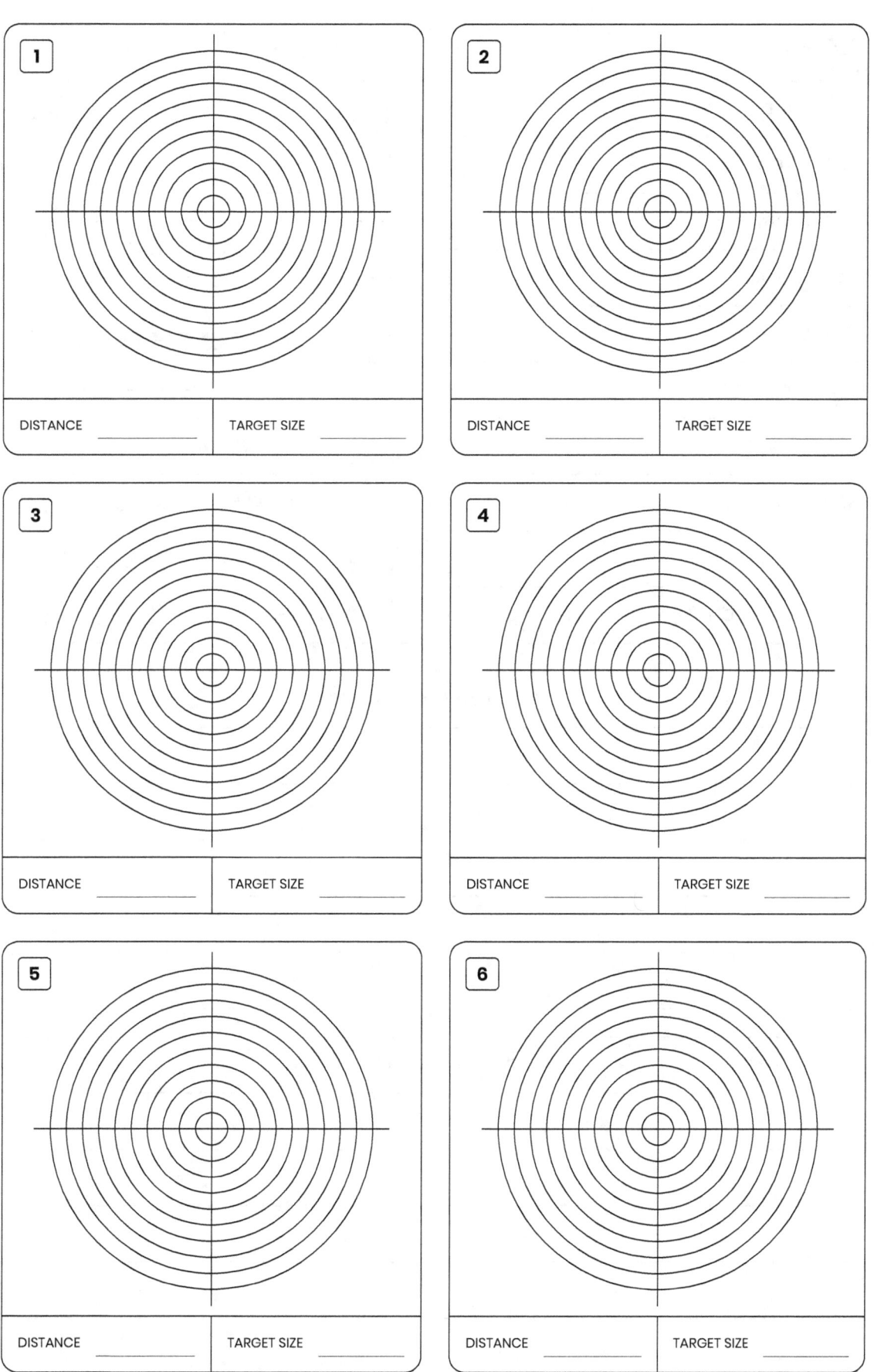

📅 DATE	
🕐 TIME	
📍 LOCATION	
👥 PARTNER	
🔫 FIREARM	
🎯 SCOPE TYPE	
🔩 BULLET TYPE	
⛰ POWDER	
💥 PRIMER	
✒ BRASS	
🔫 SEATING DEPTH	

WEATHER CONDITIONS

🌡 ____ ☀ ⛅ 🌧 ⛈ ❄
🚩 ____ ☐ ☐ ☐ ☐ ☐

LIGHT CONDITIONS

☀ 1 — 2 — 3 — 4 — 5 🌙
BRIGHT ○ ○ ○ ○ ○ DARK

RATING

🔫	WEAPON HANDLING	☆☆☆☆☆
🎯	HIT RATE	☆☆☆☆☆
📋	OVERALL RESULTS	☆☆☆☆☆

SESSION HIGHLIGHTS

ADDITIONAL NOTES

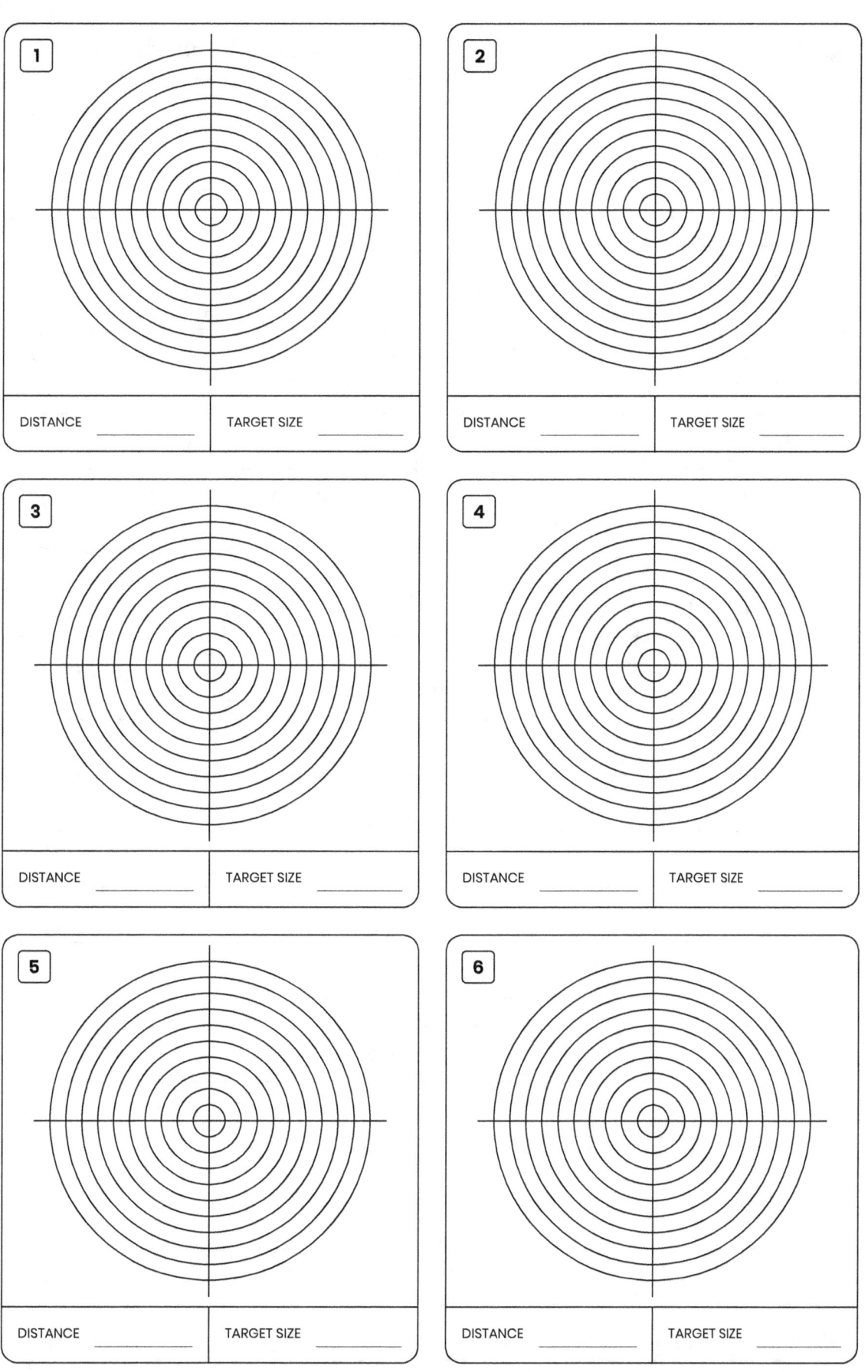

📅	**DATE**
🕐	**TIME**
📍	**LOCATION**
👥	**PARTNER**
🔫	**FIREARM**
🎯	**SCOPE TYPE**
🔹	**BULLET TYPE**
	POWDER
	PRIMER
	BRASS
	SEATING DEPTH

WEATHER CONDITIONS

🌡 ____   ☀ ⛅ 🌧 ⛈ ❄
🚩 ____   ☐ ☐ ☐ ☐ ☐

LIGHT CONDITIONS

☀ 1 — 2 — 3 — 4 — 5 🌙
BRIGHT ○ ○ ○ ○ ○ DARK

RATING

🔫	WEAPON HANDLING	☆☆☆☆☆
🎯	HIT RATE	☆☆☆☆☆
📋	OVERALL RESULTS	☆☆☆☆☆

SESSION HIGHLIGHTS

___

ADDITIONAL NOTES

___

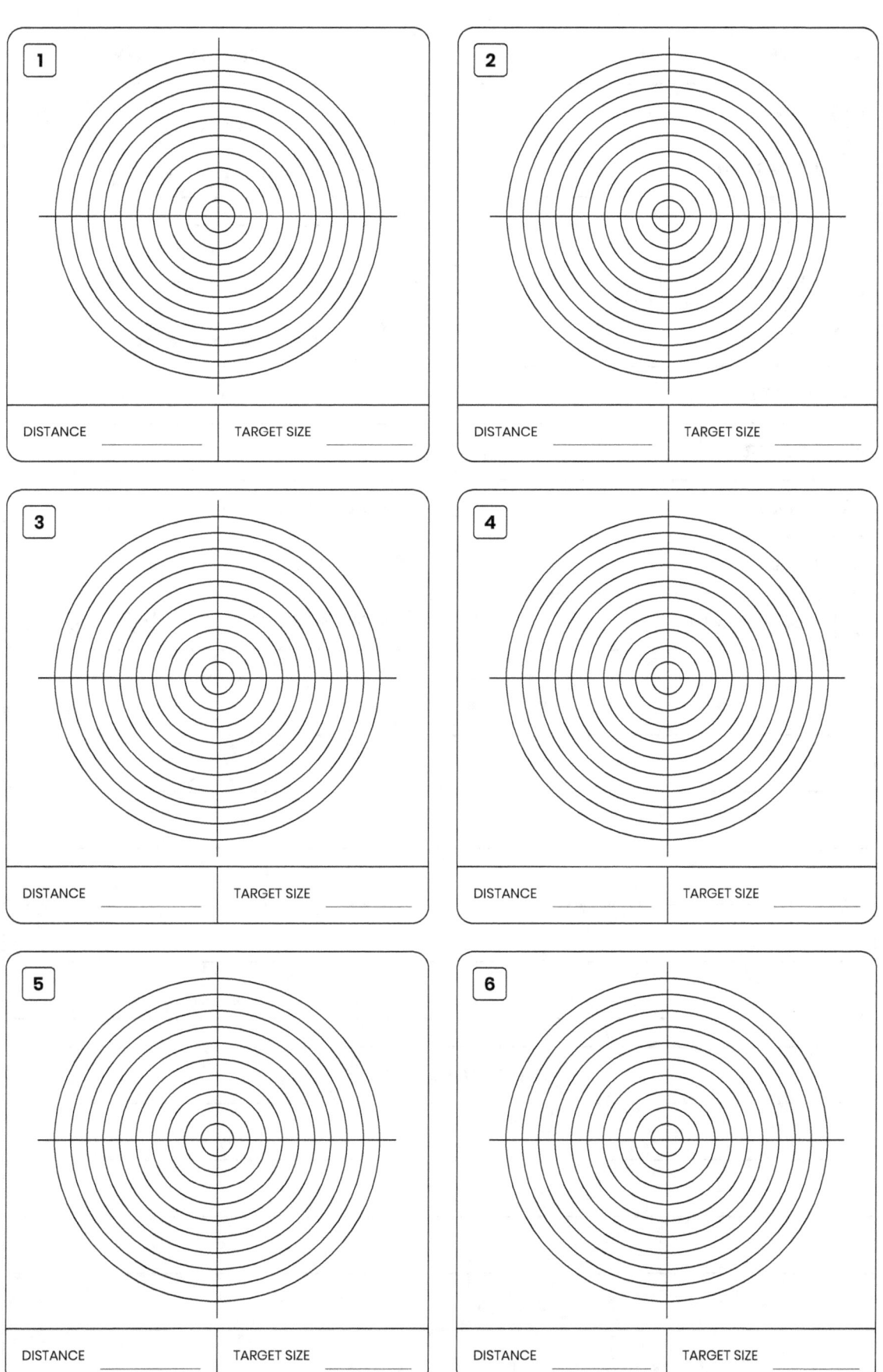

📅	**DATE**
🕐	**TIME**
📍	**LOCATION**
👥	**PARTNER**
🔫	**FIREARM**
🎯	**SCOPE TYPE**
	BULLET TYPE
	POWDER
	PRIMER
	BRASS
	SEATING DEPTH

WEATHER CONDITIONS

🌡️ ___ ☀️ ⛅ 🌧️ ⛈️ ❄️

🚩 ___ ☐ ☐ ☐ ☐ ☐

LIGHT CONDITIONS

BRIGHT 1 — 2 — 3 — 4 — 5 DARK

RATING

	WEAPON HANDLING	☆☆☆☆☆
	HIT RATE	☆☆☆☆☆
	OVERALL RESULTS	☆☆☆☆☆

SESSION HIGHLIGHTS

ADDITIONAL NOTES

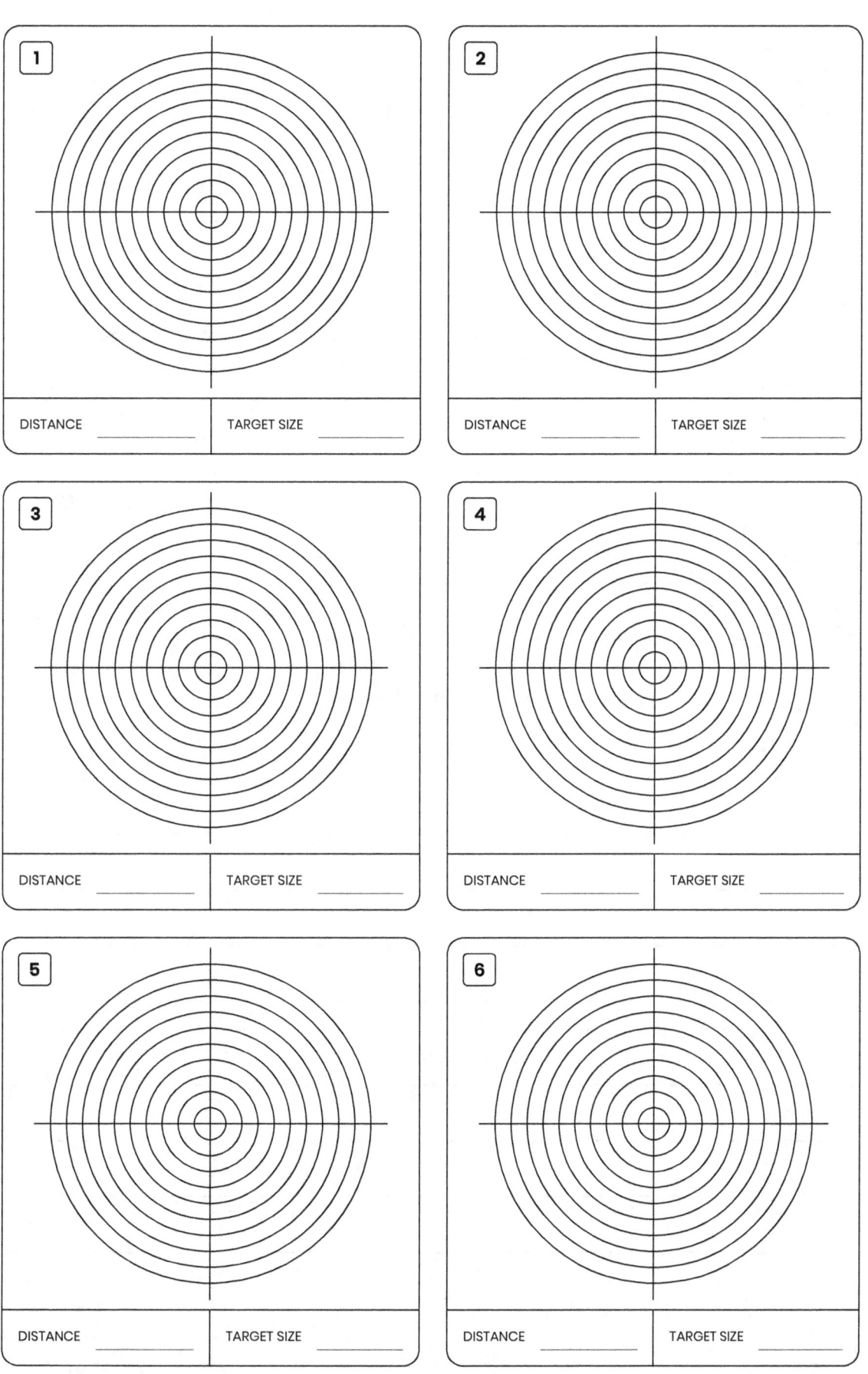

📅	**DATE**
🕐	**TIME**
📍	**LOCATION**
👥	**PARTNER**
🔫	**FIREARM**
🎯	**SCOPE TYPE**
🔘	**BULLET TYPE**
	POWDER
	PRIMER
	BRASS
	SEATING DEPTH

WEATHER CONDITIONS

🌡 _____   ☀️ ⛅ 🌧 ⛈ ❄️
🚩 _____   ☐ ☐ ☐ ☐ ☐

LIGHT CONDITIONS

☀️ 1 2 3 4 5 🌙
BRIGHT ○ ○ ○ ○ ○ DARK

RATING

	WEAPON HANDLING	☆☆☆☆☆
	HIT RATE	☆☆☆☆☆
	OVERALL RESULTS	☆☆☆☆☆

SESSION HIGHLIGHTS

ADDITIONAL NOTES

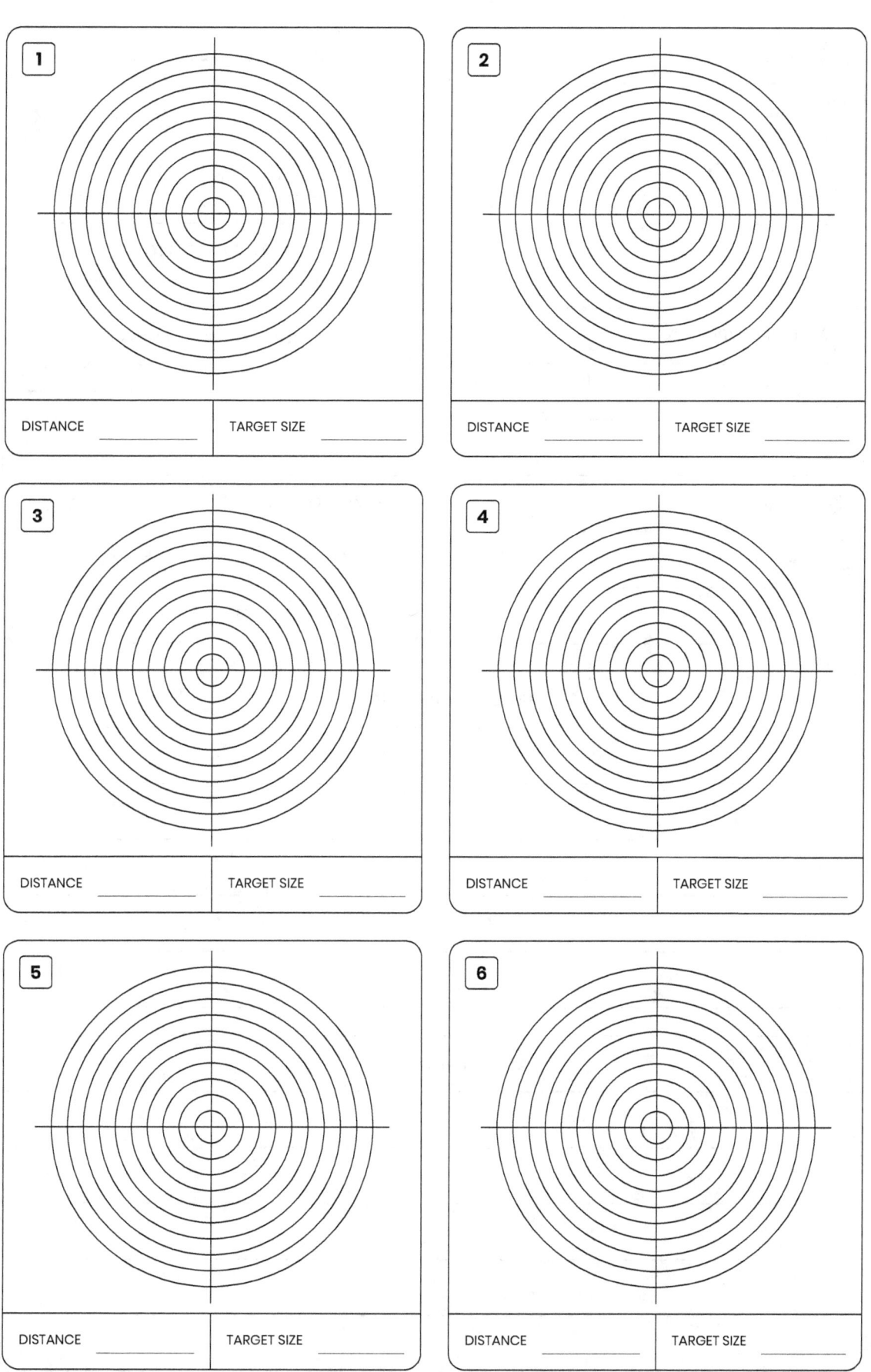

📅	**DATE**
🕐	**TIME**
📍	**LOCATION**
👥	**PARTNER**
🔫	**FIREARM**
🎯	**SCOPE TYPE**
🔩	**BULLET TYPE**
	POWDER
	PRIMER
	BRASS
	SEATING DEPTH

WEATHER CONDITIONS

🌡 _____ ☀️ ⛅ 🌧 ⛈ ❄️

🚩 _____ ☐ ☐ ☐ ☐ ☐

LIGHT CONDITIONS

☀️ 1 — 2 — 3 — 4 — 5 🌙

BRIGHT   DARK

RATING

	WEAPON HANDLING	☆☆☆☆☆
	HIT RATE	☆☆☆☆☆
	OVERALL RESULTS	☆☆☆☆☆

SESSION HIGHLIGHTS

ADDITIONAL NOTES

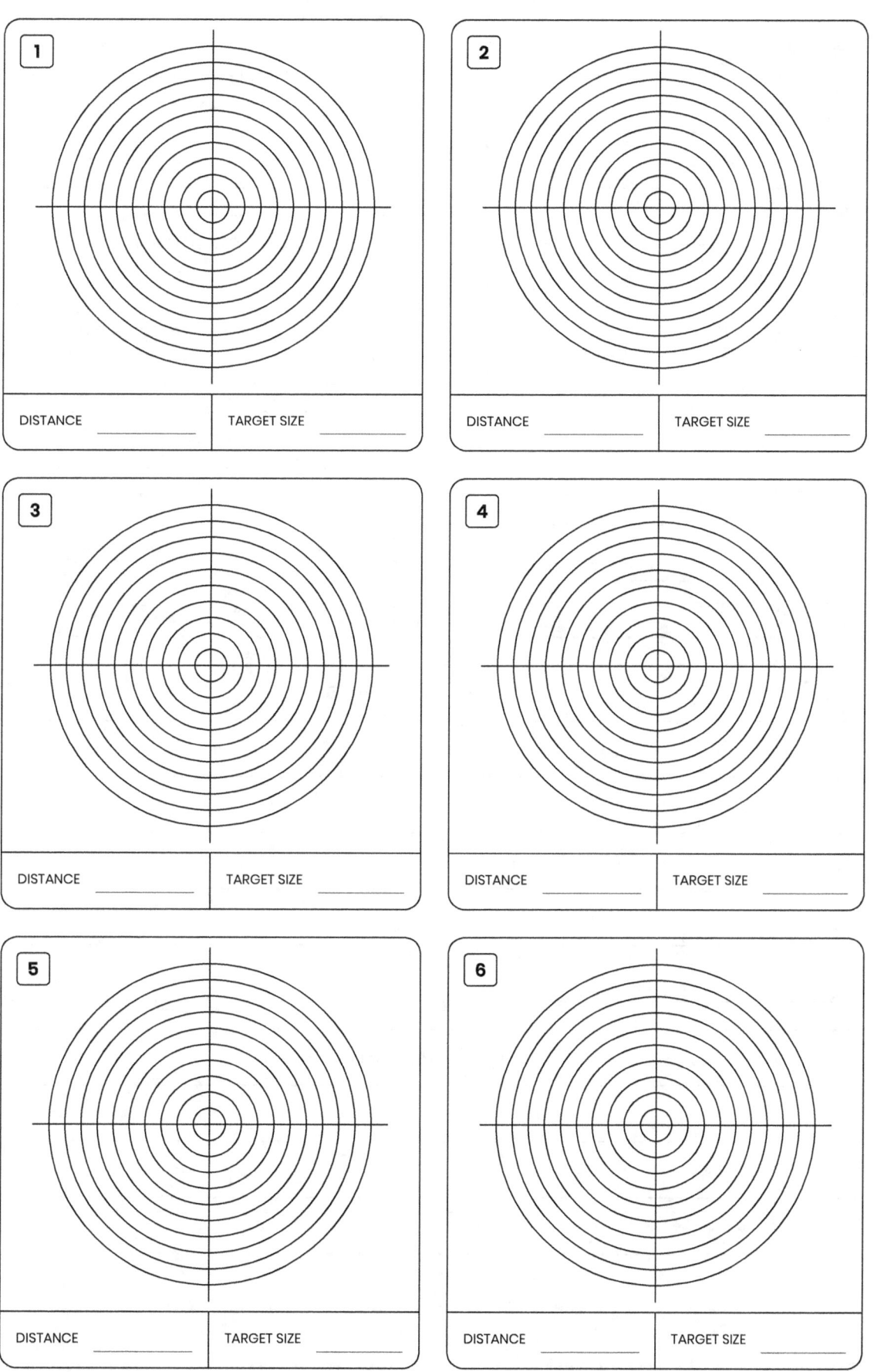

📅	DATE
🕐	TIME
📍	LOCATION
👥	PARTNER
🔫	FIREARM
◎	SCOPE TYPE
🔸	BULLET TYPE
⛰	POWDER
💥	PRIMER
🔹	BRASS
🔫	SEATING DEPTH

WEATHER CONDITIONS

🌡 _____ ☀ ⛅ 🌧 ⛈ ❄
🚩 _____ ☐ ☐ ☐ ☐ ☐

LIGHT CONDITIONS

☀ 1 2 3 4 5 🌙
BRIGHT ○ ○ ○ ○ ○ DARK

RATING

🔫	WEAPON HANDLING	☆☆☆☆☆
🎯	HIT RATE	☆☆☆☆☆
📋	OVERALL RESULTS	☆☆☆☆☆

SESSION HIGHLIGHTS

_____
_____
_____
_____
_____

ADDITIONAL NOTES

_____
_____
_____
_____
_____

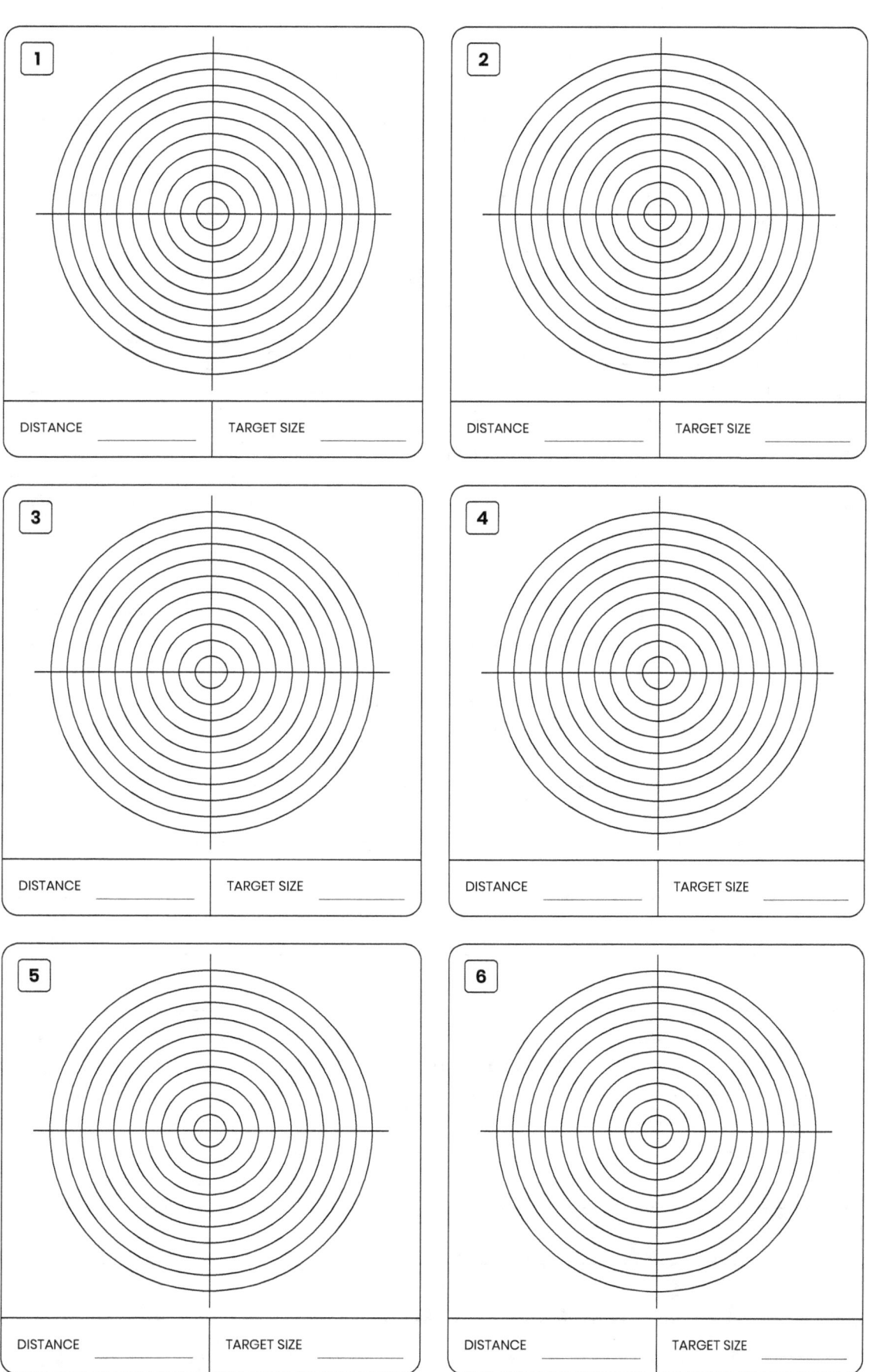

📅	DATE
🕐	TIME
📍	LOCATION
👥	PARTNER
🔫	FIREARM
🎯	SCOPE TYPE
🔸	BULLET TYPE
	POWDER
	PRIMER
	BRASS
	SEATING DEPTH

WEATHER CONDITIONS

🌡 ____   ☀️ ⛅ 🌧 ⛈ ❄️
🚩 ____   ☐ ☐ ☐ ☐ ☐

LIGHT CONDITIONS

☀️ 1 — 2 — 3 — 4 — 5 🌙
BRIGHT ○ ○ ○ ○ ○ DARK

RATING

🔫	WEAPON HANDLING	☆☆☆☆☆
🎯	HIT RATE	☆☆☆☆☆
📋	OVERALL RESULTS	☆☆☆☆☆

SESSION HIGHLIGHTS

_____
_____
_____
_____
_____

ADDITIONAL NOTES

_____
_____
_____
_____
_____

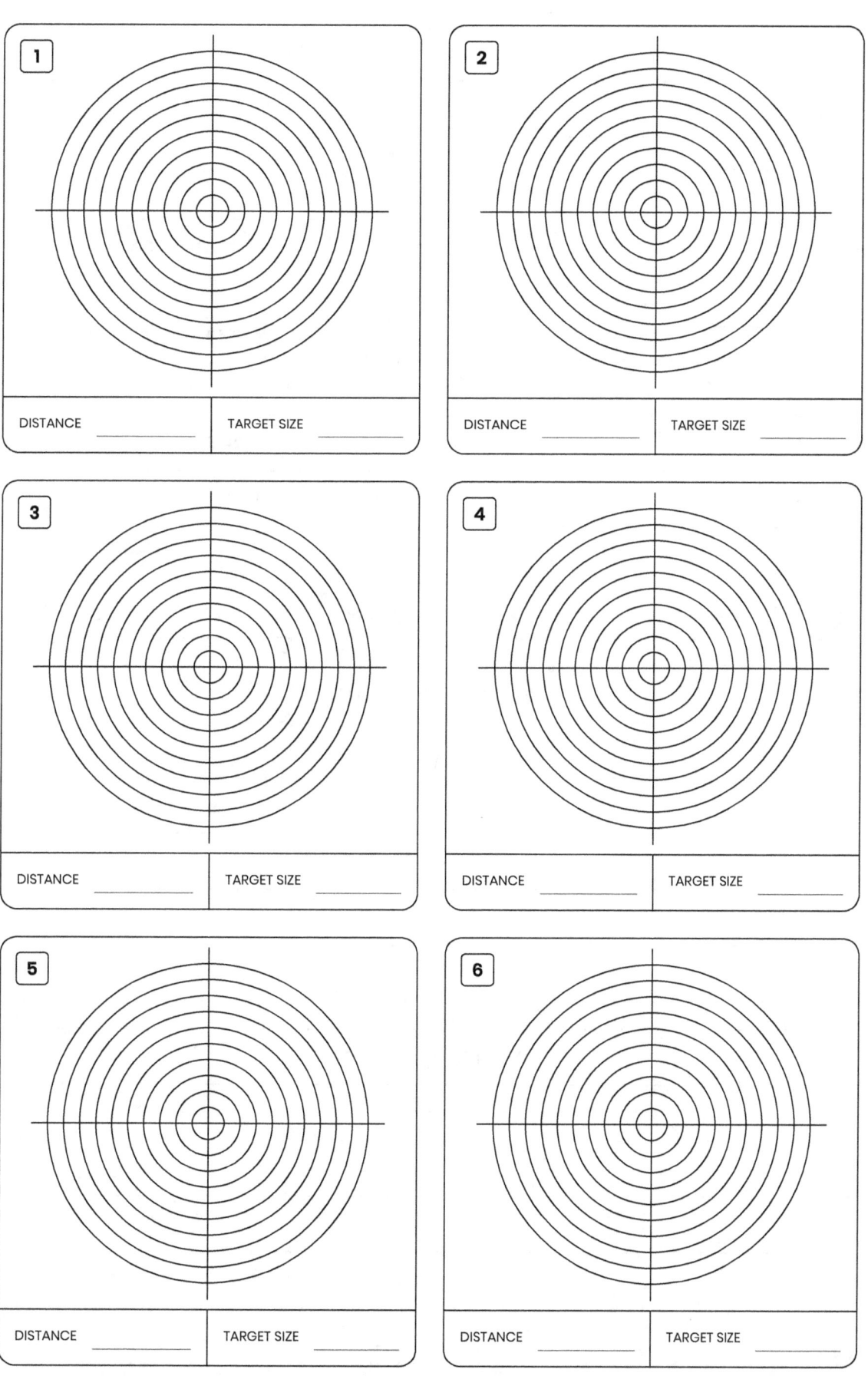

📅	**DATE**
🕐	**TIME**
📍	**LOCATION**
👥	**PARTNER**
🔫	**FIREARM**
🎯	**SCOPE TYPE**
🔸	**BULLET TYPE**
	POWDER
	PRIMER
	BRASS
	SEATING DEPTH

WEATHER CONDITIONS

🌡 ____ ☀️ ⛅ 🌧 ⛈ ❄️

🚩 ____ ☐ ☐ ☐ ☐ ☐

LIGHT CONDITIONS

☀️ 1 2 3 4 5 🌙
BRIGHT ○ ○ ○ ○ ○ DARK

RATING

🔫	WEAPON HANDLING	☆☆☆☆☆
🎯	HIT RATE	☆☆☆☆☆
📋	OVERALL RESULTS	☆☆☆☆☆

SESSION HIGHLIGHTS

_____
_____
_____
_____
_____

ADDITIONAL NOTES

_____
_____
_____
_____
_____

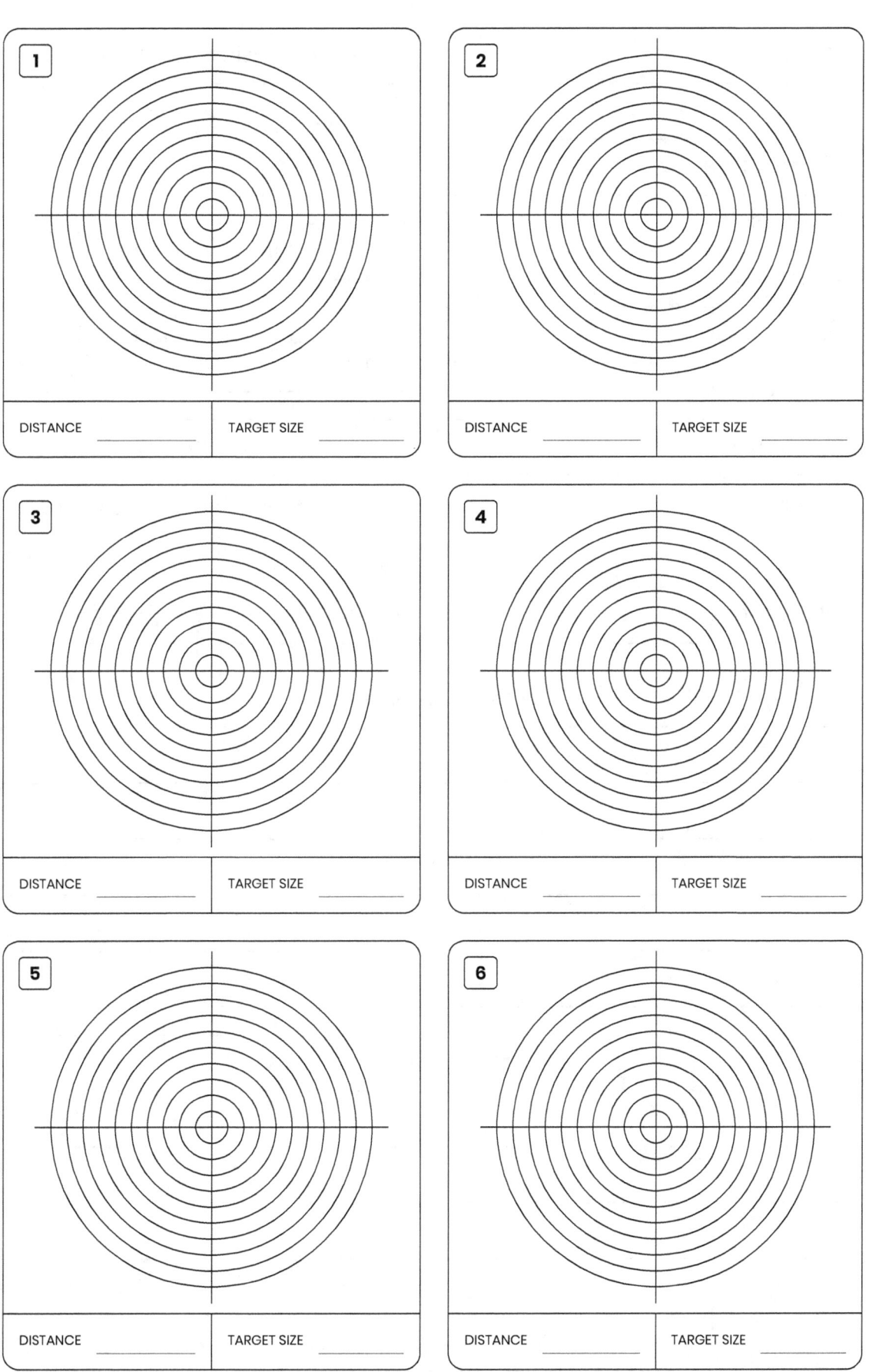

📅	**DATE**
🕐	**TIME**
📍	**LOCATION**
👥	**PARTNER**
🔫	**FIREARM**
🎯	**SCOPE TYPE**
🔹	**BULLET TYPE**
	POWDER
	PRIMER
	BRASS
	SEATING DEPTH

WEATHER CONDITIONS

🌡 ____   ☀ ☁ 🌧 ⛈ ❄
🚩 ____   ☐ ☐ ☐ ☐ ☐

LIGHT CONDITIONS

☀ BRIGHT —— 1 —— 2 —— 3 —— 4 —— 5 —— 🌙 DARK

RATING

	WEAPON HANDLING	☆☆☆☆☆
	HIT RATE	☆☆☆☆☆
	OVERALL RESULTS	☆☆☆☆☆

SESSION HIGHLIGHTS

ADDITIONAL NOTES

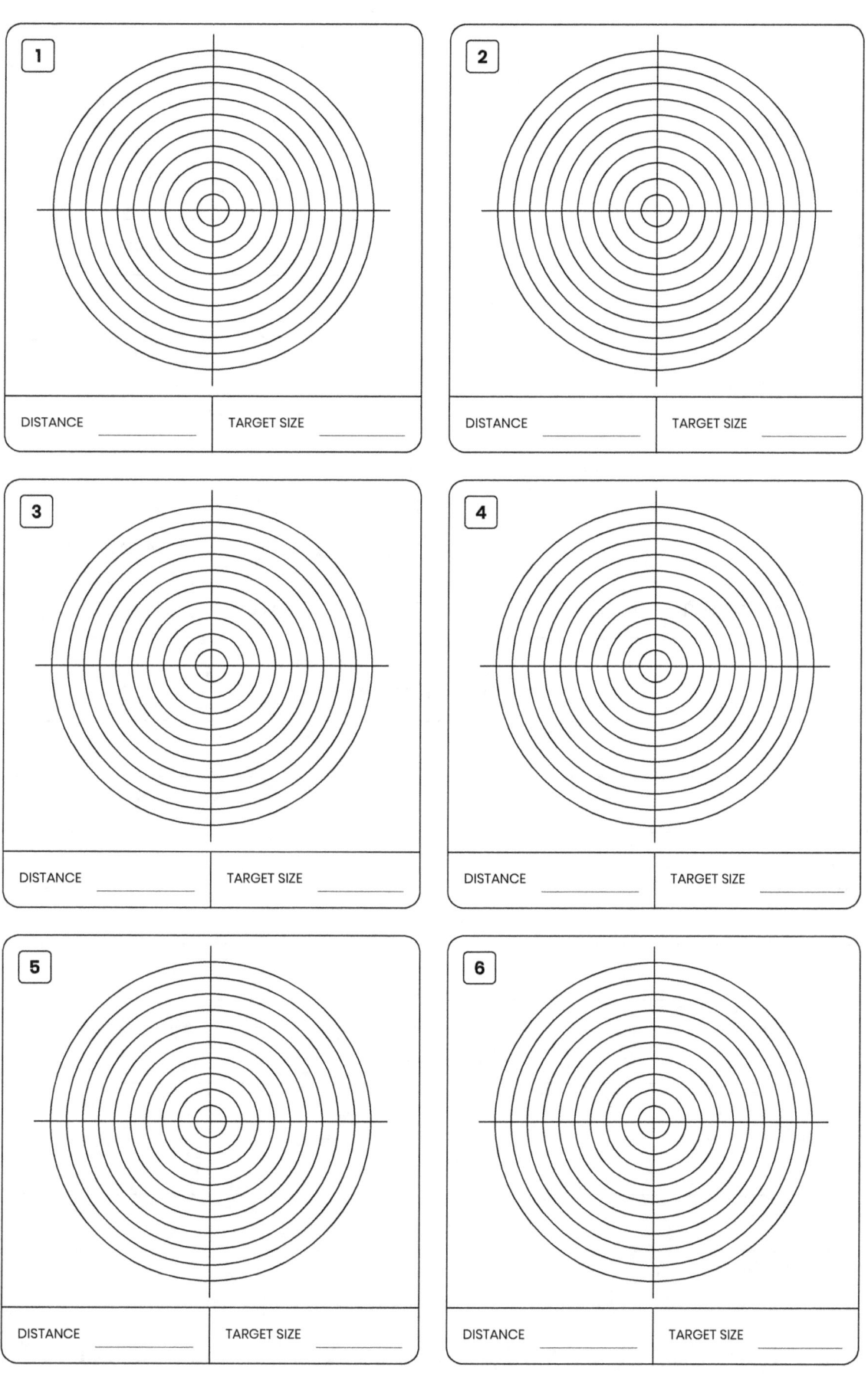

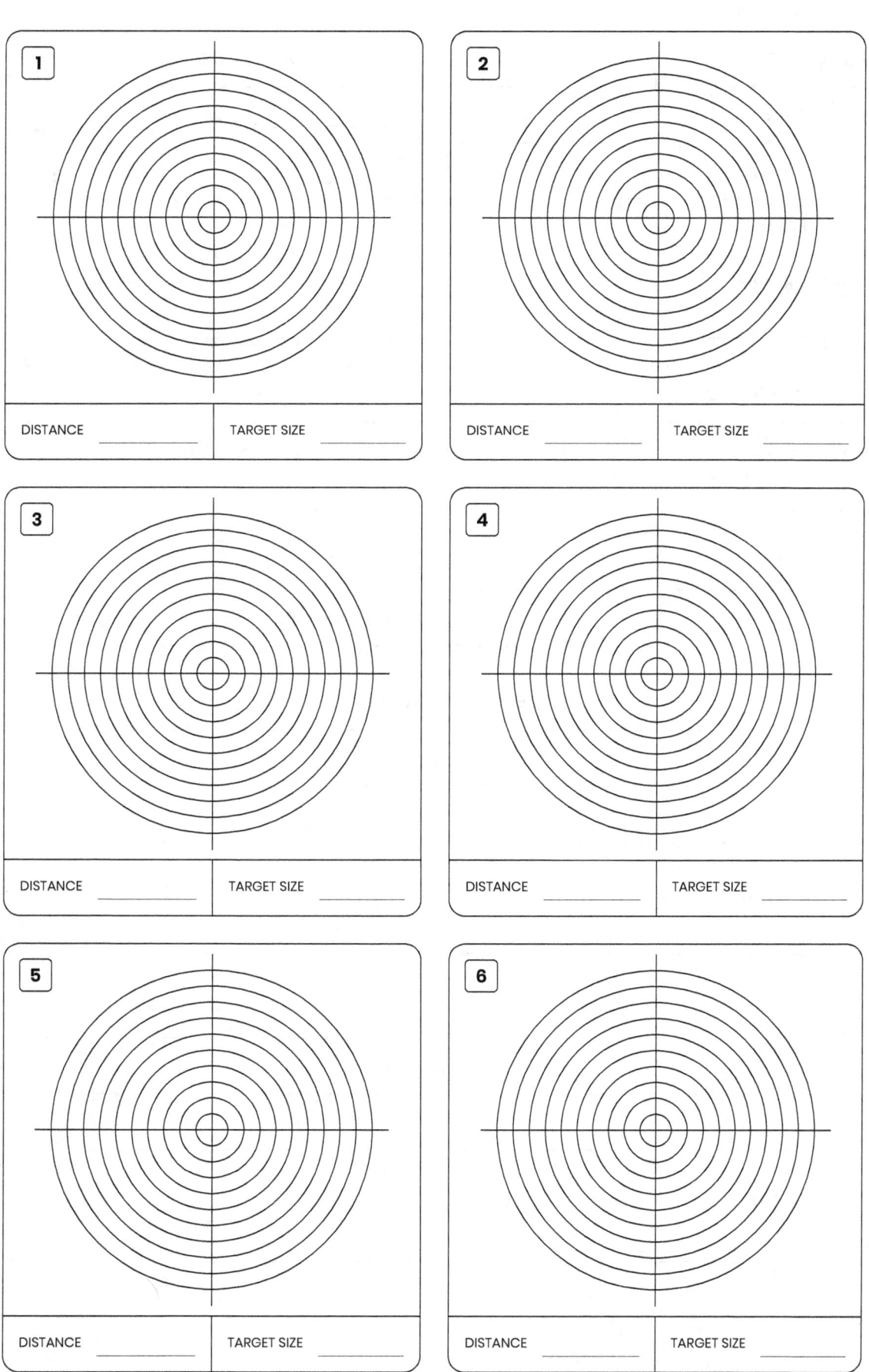

📅	**DATE**
🕐	**TIME**
📍	**LOCATION**
👥	**PARTNER**
🔫	**FIREARM**
🎯	**SCOPE TYPE**
🔹	**BULLET TYPE**
	POWDER
	PRIMER
	BRASS
	SEATING DEPTH

WEATHER CONDITIONS

🌡 ___ ☀️ ⛅ 🌧 ⛈ ❄️
🚩 ___ ☐ ☐ ☐ ☐ ☐

LIGHT CONDITIONS

☀️ 1 — 2 — 3 — 4 — 5 🌙
BRIGHT ○ ○ ○ ○ ○ DARK

RATING

	WEAPON HANDLING	☆☆☆☆☆
	HIT RATE	☆☆☆☆☆
	OVERALL RESULTS	☆☆☆☆☆

SESSION HIGHLIGHTS

ADDITIONAL NOTES

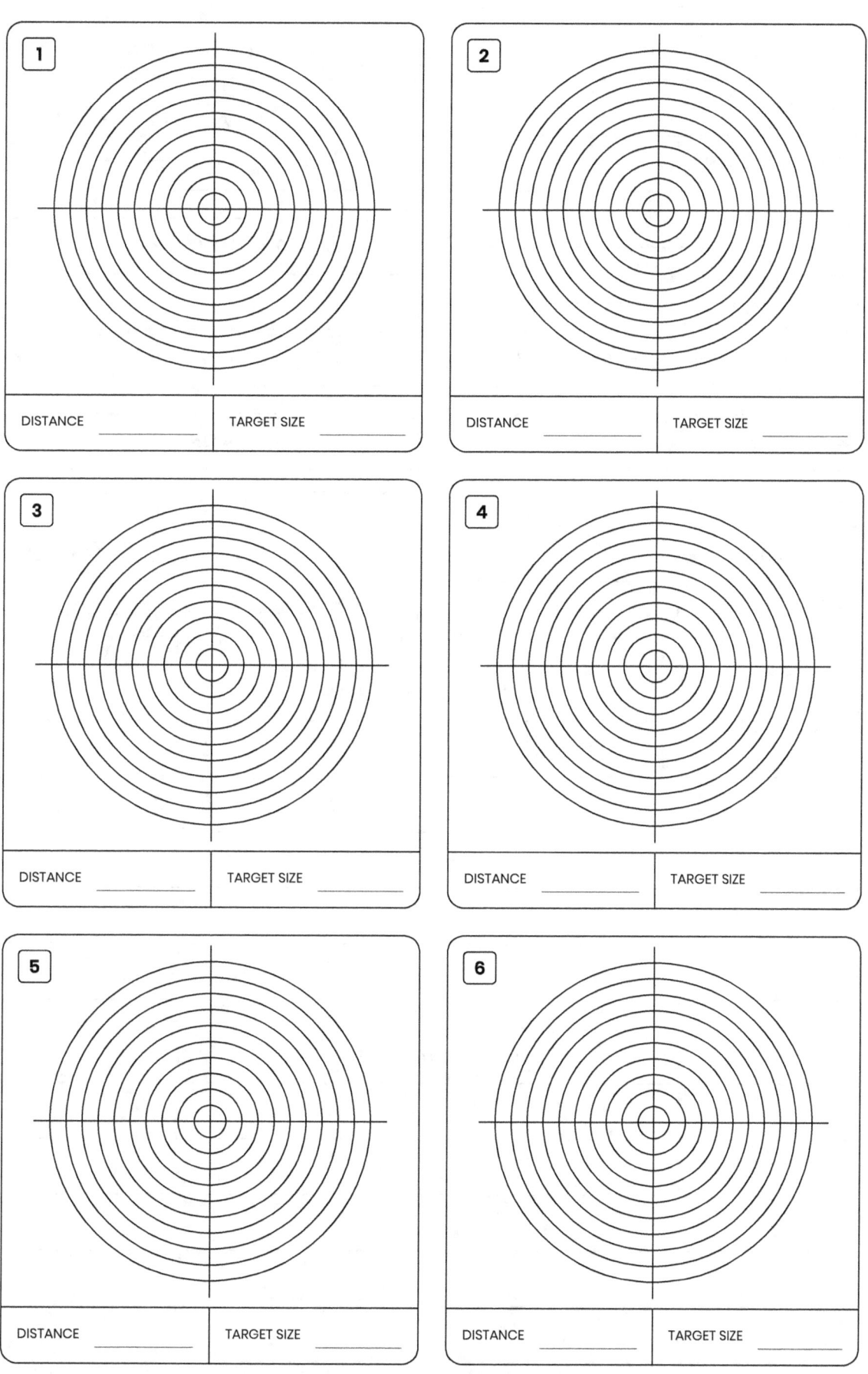

📅	DATE
🕐	TIME
📍	LOCATION
👥	PARTNER
🔫	FIREARM
⊙	SCOPE TYPE
🔸	BULLET TYPE
◇	POWDER
✨	PRIMER
🔹	BRASS
🔫	SEATING DEPTH

WEATHER CONDITIONS

🌡 ____   ☀ ⛅ 🌧 ⛈ ❄
🚩 ____   ☐ ☐ ☐ ☐ ☐

LIGHT CONDITIONS

☀ —1—2—3—4—5— 🌙
BRIGHT ○ ○ ○ ○ ○ DARK

RATING

🔫	WEAPON HANDLING	☆☆☆☆☆
🎯	HIT RATE	☆☆☆☆☆
📋	OVERALL RESULTS	☆☆☆☆☆

SESSION HIGHLIGHTS

_____
_____
_____
_____
_____

ADDITIONAL NOTES

_____
_____
_____
_____
_____

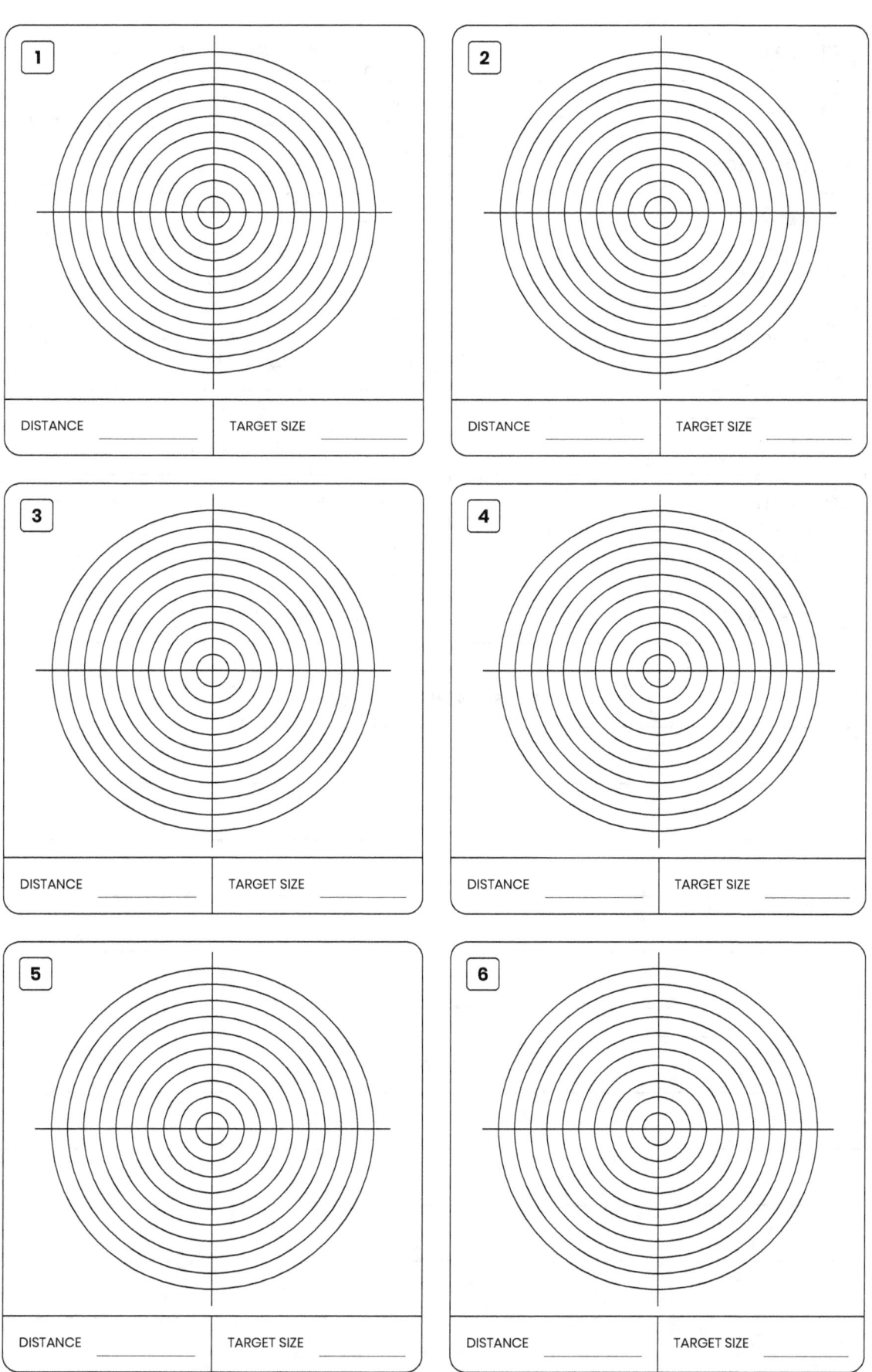

📅	**DATE**
🕐	**TIME**
📍	**LOCATION**
👥	**PARTNER**
🔫	**FIREARM**
🎯	**SCOPE TYPE**
	BULLET TYPE
	POWDER
	PRIMER
	BRASS
	SEATING DEPTH

WEATHER CONDITIONS

🌡️ _____   ☀️ ⛅ 🌧️ ⛈️ ❄️
🚩 _____   ☐   ☐   ☐   ☐   ☐

LIGHT CONDITIONS

☀️ 1   2   3   4   5 🌙
BRIGHT ○   ○   ○   ○   ○ DARK

RATING

🔫	WEAPON HANDLING	☆☆☆☆☆
🎯	HIT RATE	☆☆☆☆☆
📋	OVERALL RESULTS	☆☆☆☆☆

SESSION HIGHLIGHTS

_____
_____
_____
_____
_____

ADDITIONAL NOTES

_____
_____
_____
_____
_____

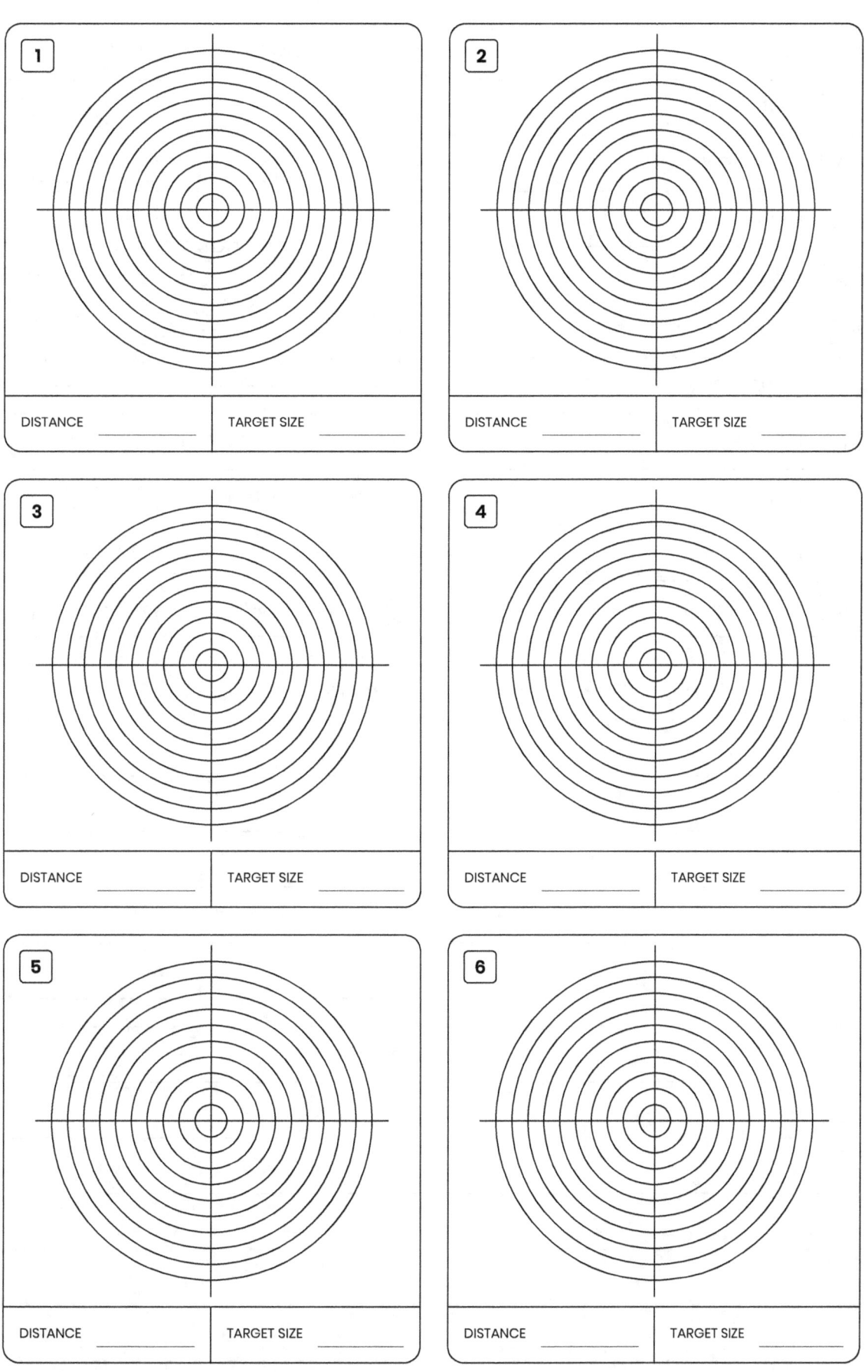

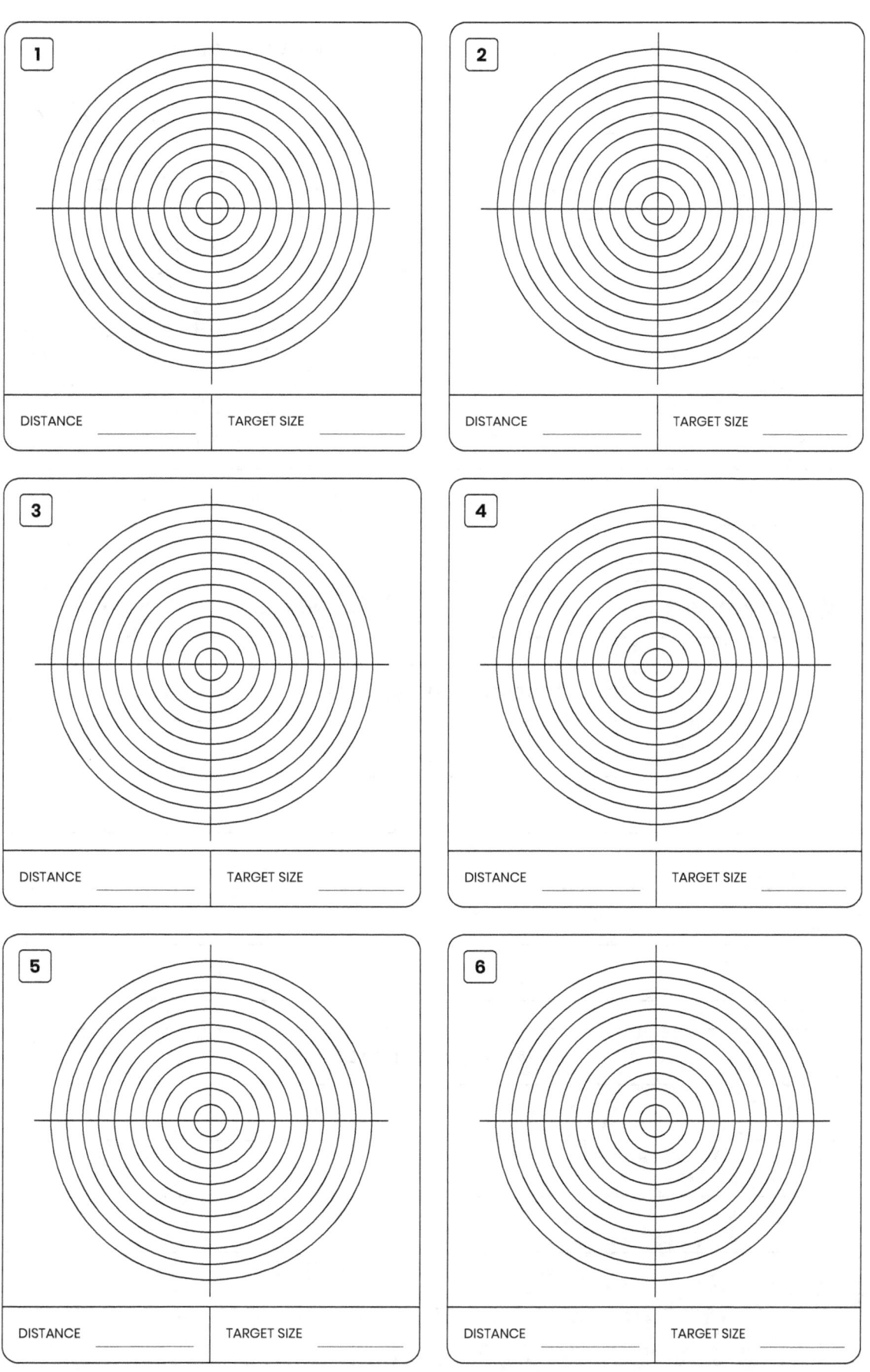

📅	**DATE**
🕐	**TIME**
📍	**LOCATION**
👥	**PARTNER**
🔫	**FIREARM**
◎	**SCOPE TYPE**
🔹	**BULLET TYPE**
	POWDER
	PRIMER
	BRASS
	SEATING DEPTH

WEATHER CONDITIONS

🌡 ____ ☀️ ⛅ 🌧 ⛈ ❄️
🚩 ____ ☐ ☐ ☐ ☐ ☐

LIGHT CONDITIONS

☀️ 1 — 2 — 3 — 4 — 5 🌙
BRIGHT ○ ○ ○ ○ ○ DARK

RATING

	WEAPON HANDLING	☆☆☆☆☆
	HIT RATE	☆☆☆☆☆
	OVERALL RESULTS	☆☆☆☆☆

SESSION HIGHLIGHTS

_____
_____
_____
_____
_____

ADDITIONAL NOTES

_____
_____
_____
_____
_____

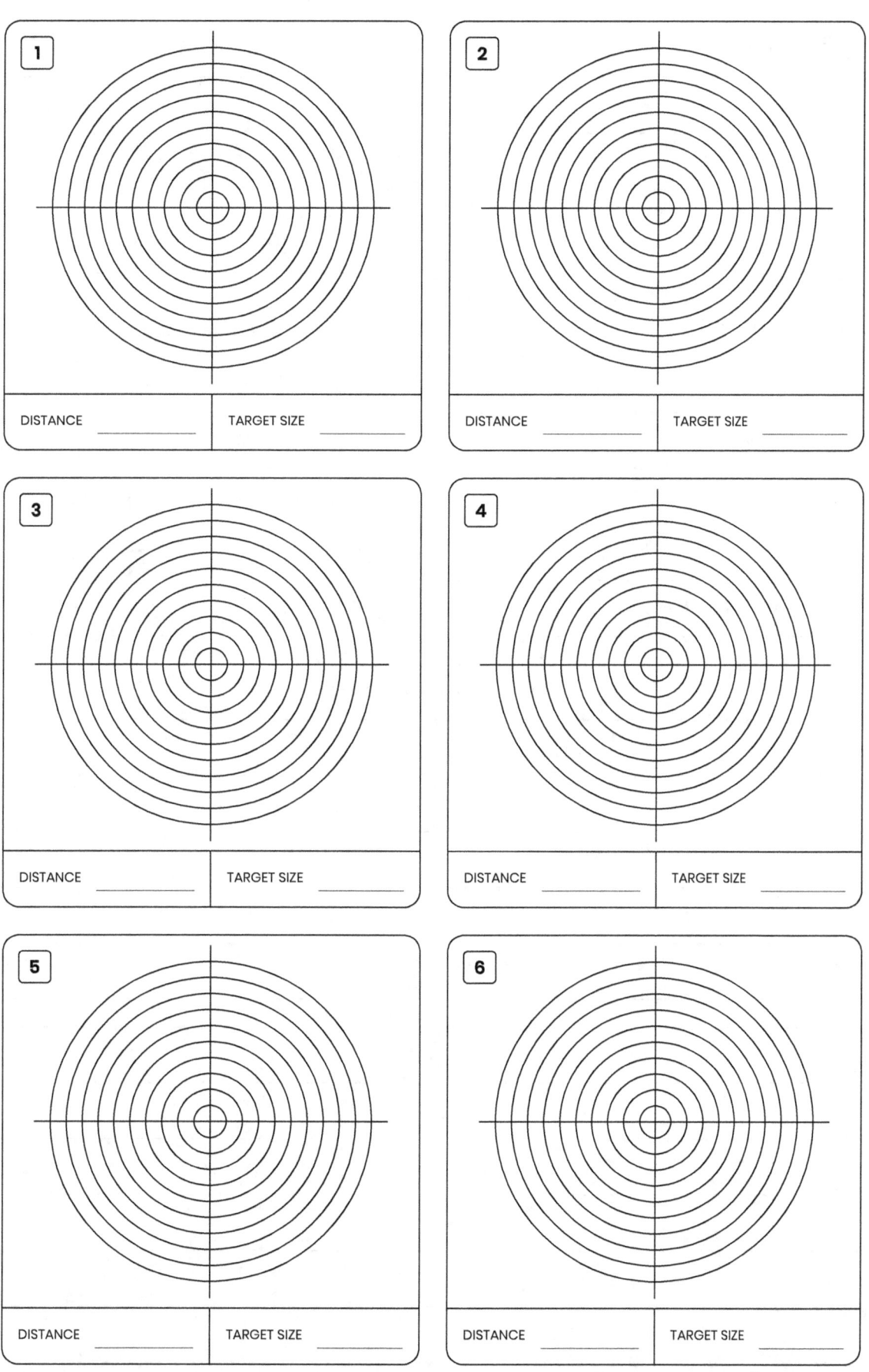

DATE

TIME

LOCATION

PARTNER

FIREARM

SCOPE TYPE

BULLET TYPE

POWDER

PRIMER

BRASS

SEATING DEPTH

WEATHER CONDITIONS

LIGHT CONDITIONS

1 2 3 4 5
BRIGHT — DARK

RATING

- WEAPON HANDLING ☆☆☆☆☆
- HIT RATE ☆☆☆☆☆
- OVERALL RESULTS ☆☆☆☆☆

SESSION HIGHLIGHTS

ADDITIONAL NOTES

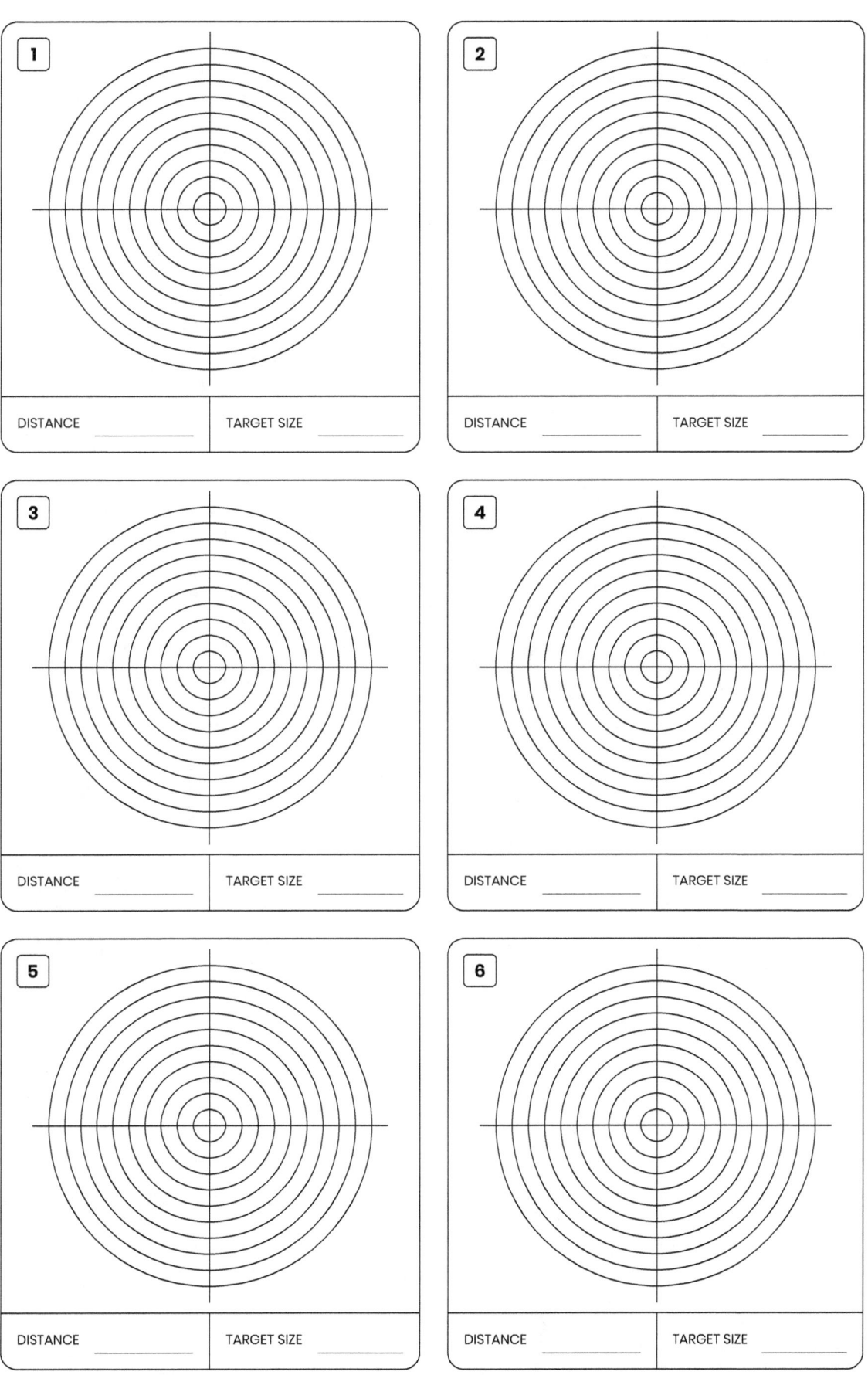

www.ingramcontent.com/pod-product-compliance
Lightning Source LLC
Chambersburg PA
CBHW081154070526
44583CB00021B/2839